The Art and Science of D.... ...
Wood-Fired Ovens

The Art and Science of Dome-Shaped Wood-Fired Ovens, from history to your backyard or commercial shop, is a carefully crafted guide that explains the tradition and science of wood-fired cooking. The book embarks on a historical journey, tracing the development of wood-fired ovens and their cultural significance. It then unravels the theory of heating and the burning behaviour of wood, making complex technical concepts accessible.

Transitioning from theory to practice, the guide outlines the design and construction process of a wood-fired oven. It considers engineering aspects and locally available materials, emphasizing efficient, sustainable building. The book discusses essential cooking utensils and tools, shedding light on the entire cooking process, from fire-starting to ash disposal.

In a unique chapter on data logging, readers are introduced to modern temperature monitoring techniques. It shows how managing thermal mass can expand the range of recipes beyond the commonly perceived breads and pizzas. Lastly, the book explores Turkish cuisine, debunking preconceptions and presenting a wide array of dishes suitable for wood-fired ovens. The recipes span from traditional Turkish to International cuisines and fusion recipes, equipping readers with the tools to broaden their culinary repertoire.

This book serves as an indispensable resource for anyone interested in wood-fired cooking, blending historical context, technical insights, practical advice, and mouth-watering recipes into a compelling narrative. This comprehensive manual aims to bring wood-fired cooking into the heart of modern culinary practice.

The Art and Science of Dome-Shaped Wood-Fired Ovens
Theory, Building Techniques, Thermal Profiling

Nesimi Ertuğrul

*With Traditional Recipes
and Fusion Cooking*

CRC Press
Taylor & Francis Group
Boca Raton London New York

CRC Press is an imprint of the
Taylor & Francis Group, an **informa** business

First edition published 2024
by CRC Press
2385 NW Executive Center Drive, Suite 320, Boca Raton FL 33431

and by CRC Press
4 Park Square, Milton Park, Abingdon, Oxon, OX14 4RN

CRC Press is an imprint of Taylor & Francis Group, LLC

ISBN: 978-1-032-64012-9 (hbk)
ISBN: 978-1-032-62623-9 (pbk)
ISBN: 978-1-032-64013-6 (ebk)

DOI: 10.1201/9781032640136

Typeset in Times New Roman
by MPS Limited, Dehradun

Access the recipe website: http://woodfirewonders.com/book-recipes/

This work is dedicated to

my late father, Ahmet, who left an indelible imprint of the taste and aroma of wood-fired oven-cooked ribs (buryan) in my memory,

my dearests, Olivia and Atacan, who have grown up with the heartwarming tradition of the wood-fired oven,

and "The Restaurant of Mistaken Orders" in Tokyo.

Contents

Diet and love-making, all primal needs of every human being

Confucius

Wood-fired oven cooking takes us back to basics, connecting us with ancestral cooking methods, and it infuses dishes with a richness of culinary experiences and represents one of the healthiest forms of cooking.

Foreword

"The Oven is the Heart of Kitchen"

In Kelkit Town, near our ancestral home, just 100 metres away, there was a stone oven house built by my grandfather, İrfani Doğan, in the 1940s on an empty plot of land in front of Faik Bey Uncle's house. Constructed entirely of natural stone and clay, this environmentally friendly oven house was expertly and wisely built. Inside the house, a stone oven was placed on an elevated pedestal. My generous and philanthropic grandfather had this oven house built to serve the entire town. The placement of firewood and the cleanliness of the oven were under the control of our diligent servant, Güller Bibi*. The women of the neighbourhood would discuss and arrange a schedule among themselves for bread-making.

As a child in the 60s, I loved entering this oven house and indulging in freshly baked bread, still steaming with the smoke of the stone oven. Later, I saw that a small room made of clay, goat hair, and straw mixture was built adjacent to the oven house. Additionally, a system was set up to transfer the heat from the oven house to this small room, creating a miniature bathhouse with hot water flowing from its fountain. During the freezing cold winter, children would bathe in this miniature bathhouse. I learned from my Aunt Nuran that the remains of the small bathhouse within the abandoned oven can still be seen today.

Stone ovens, a traditional cooking method used since ancient times, hold great importance in Turkish culinary culture. Traditionally, bread, pide, and simit, as well as meat and herb-filled thin breads, are all baked in stone ovens. In the Black Sea region, stone ovens are commonly referred to as "kara fırın" (black oven) due to the soot and darkening caused by the wood fire on the inner walls of the oven. The heating section of the oven, where the firewood is burned, is usually located on the left side, while bread and dishes are cooked on the right side. This section is known as the "koltuk kısmı" (seat part) or the guest room of the oven. This is because bread goes in, bakes, and comes out, followed by pastries and then casseroles. It's a continuous flow of baking. In the Southeastern Anatolia region, the secret to the deliciousness of tray dishes with meat and vegetables, as well as casserole dishes cooked in clay pots, lies in the stone oven.

Archaeological excavations have revealed the existence of cooking stoves in Anatolian residential architecture since the Neolithic period, which dates back to around 7400 BC. The stoves, designed in different shapes according to their purpose, were generally portable or built adjacent to walls. Çatalhöyük, one of Anatolia's first settlement cities with a history dating back to 7400 BC, had rectangular stoves placed in the centre of houses. The Hittites, one of Anatolia's oldest civilizations, who lived around the 17th century BC, constructed wall-leaning horseshoe-shaped stoves. Contemporary with the stoves, the ovens found were planned with domed roofs, made of stone halfway up and clayey soil on top.

Furthermore, archaeological excavations have revealed the presence of tandır ovens alongside stoves and ovens in Eastern Anatolian house architecture. Tandır ovens, used for both cooking and heating purposes since prehistoric times, are prepared by digging a pit in the ground. These enclosed ovens, round in shape and lined with clayey soil, reach a depth of up to 150 cm with a mouth opening of about 55–65 cm. They are also referred to as "tennur" ovens. Although not as common, in some villages, stone ovens and tandır ovens continue to peacefully and harmoniously serve to satiate people's appetites even today.

The Art and Science of Dome-Shaped Wood-Fired Ovens deserves recognition for its comprehensive exploration of wood-fired ovens. While these ovens are commonly found, their technical designs and performance have often been overlooked in terms of modern analysis. Ertuğrul's book fills this gap by providing a scientifically rigorous examination of the history, design, and construction of wood-burning ovens. The book also delves into the various types of heat transfers using simplified analogies, dispelling misconceptions that wood-fired ovens are solely for baking bread. I strongly believe that this book will fill the gap and serve as a valuable resource for anyone interested in understanding and mastering the art of wood-fired oven cooking.

Sahrap Soysal
Istanbul, 19 June 2023

Sahrap Soysal is a prominent cookbook author and culinary figure in Turkiye. She gained recognition by preparing and hosting the programme "Mutfakta Keyif" (Joy in the Kitchen) starting in 2001. Soysal has authored several unique cookbooks with a strong emphasis on the traditional and cultural aspects of Turkish cuisine. Her book "Bir Yemek Masalı" (A Food Tale) received the prestigious Gourmand Best Cookbook in the World Award in 2004, and "Sevgilim, Akşama Ne Pişirdin?" (My Love, What Did You Cook for Dinner?) was honoured as the Gourmand Cookbook of the Year in 2007. For 23 years, her column writing for *Hürriyet* has continued. Presently, she continues to host television programmes centred around culinary topics and serves as a culinary consultant and editor for newspapers and magazines.

*Bibi: An affectionate term used to refer to an older woman or grandmother, similar to the "granny" or "nana." It is also a familial term used to show respect and closeness to a female relative. In addition, it is used to refer to an aunt, specifically a father's sister, in Turkish.

Preface

It's widely acknowledged that babies experience smell and taste before they acquire language. They use their mouths to explore and learn different textures and tastes, deciding whether substances are suitable for consumption or not. Hence, they taste everything they find around them, such as wood, metal, soil, or sand. As they grow older, they might forget words and the names of their favourite foods, but not the tastes of those inedible substances.

Because taste is culturally modified and we all have comfort tastes, we are universally attracted to the smell of food associated with specific cooking methods. These smells are linked to our memory and our emotions. For instance, familiar and appealing smells connect us to our favourite places and comforting friendships while strengthening our bonds to the past.

As the science of food and cooking improves, medical studies are also actively investigating how to identify better links and clear correlations between our senses, neural networks, and memory in the brain. We already know that a certain section of the brain, the olfactory cortex, is involved in the processing and perception of odour (smell) and its categorization. This section of the brain is a part of the limbic system, which is also involved in our behavioural and emotional responses. Hence, strong associations are created between particular scents (distinctive smells) and experiences. This is why, years later, a scent may trigger our memory and remind us of a time or feeling from the past. We can conclude that this system in the brain clearly involves the processing of our emotions, our memory formation, and our survival instincts while connecting to our senses.

Though cooking is a traditionally prescribed chemical process, our experience with food is essentially a combination of our senses. It's perceived by our tongue, smell, texture, and temperature. However, the description of flavour is produced only when taste is combined with smell. For example, when we have a blocked nose, our perception of taste is also dulled. Furthermore, our taste through our tongue and smell through our nose are closely linked to our emotions, as both senses are connected to our involuntary nervous system. Therefore, an individually unacceptable taste or smell can upset us, while an appetizing flavour can increase the production of saliva and gastric juices to welcome that food.

In addition, recent research studies have identified a forgotten "flavour sense." Among the seven sensory attributes (taste, smell, temperature, texture, colour, appearance, and sound), the sound of food is considered as important and essential as colour and appearance. Sound can be categorized into two different types: eating sounds associated with the end-product and consumption practice (such as crispy, crunchy, and crackly), and cooking sounds that are produced during cooking methods (such as burning wood sound, sizzling, and popping). While the former is an indicator of texture and quality, the latter is uniquely and commonly experienced when we are close to the cooking platform, such as a wood-fired oven.

When we combine various ingredients to create a dish, we also construct five known tastes (sweet, sour, salty, bitter, and savoury) that are transported from our tongue to our brain. Science has already identified the mechanisms between chemical substances and taste buds (with numerous sensory cells on a tongue), resulting in flavours. However, it's not possible to identify "all combinations of flavours" in foods when these five basic tastes at varying levels of intensity are combined with two flavour-related senses (taste and smell).

I believe that wood-fired oven cooking touches all our senses highly effectively, producing diverse and miraculous flavours each time, even when we use the same ingredients and follow the same recipes repeatedly. Although this is primarily associated with the unpredictability of wood burning, the lack of knowledge about the temperature profile of a wood-fired oven is also a significant factor. Despite the cultural richness of wood-fired ovens and their adaptation to cooking for thousands of years, there hasn't been much progress in their association with scientific approaches, engineering designs, and thermal behaviour during cooking, all of which are vital for predictable and repeatable end results.

Therefore, my motivation to write this book has been elevated by the lack of technically sound information about the construction and operation of wood-fired ovens. During my frequent food trips around the world, as well as searching relevant written materials, I've found that even well-known oven-building practices are highly diverse and not scientifically sound. Thus, in this book, I've aimed to link modern oven design techniques to traditional structures that date back to early human settlements and use firewood as a source of fuel.

This book is also a by-product of my own experiences, connecting me to my childhood tastes and smells and to my heritage through the use of a wood-fired oven. My wood-fired oven building and cooking experiences over 18 years have taught me that wood-fired oven cooking opposes trendy fast-cooking practices while allowing us to eat not only with our mouth, but also with our nose, eyes, and ears. As we participate in the preparation of dough and filling freshly cooked flatbread with our own choice of ingredients, our remaining sense, "touch," also joins the cooking feast. In cooking, we may intend to define recipes using logical reasoning with vague and sometimes imprecise statements, but it appears that they are all formed primarily by our childhood, companionship, and experiences using all our senses and are catalyzed by the most natural and flavoursome way of cooking! Understanding your oven and practicing regular cooking is the key to the creation of memorable tastes. Therefore, I've aimed to highlight the links between the heating characteristics of the oven and a wide range of recipes that can be cooked.

During some of my cooking parties, I've heard my son and daughter say, "I like this smell coming from the oven"! Probably, this was the moment when the particular food taste was transferred to their olfactory cortex and registered permanently to be retrieved years later! Whether your purpose is to build a commercial or domestic wood-fired oven, I hope this book will provide answers to most of your technical and practical questions. Additionally, you'll be able to practice a wide spectrum of traditional recipes that are cooked, shared, and included in this book.

I would like to acknowledge all my friends who have borne the brunt of my culinary development, my dear friend Eyyup Bevan for the historical discussions, and my colleagues who have generously shared their technical knowledge and eye-opening comments. These have also formed the content of this book after compiling my numerous short notes, measured and analyzed data, and photos taken over many years.

I hope that "the wood-fired oven cooking" will be a heritage of our survival, as we greatly enjoy it with our loved ones.

Nesimi Ertuğrul
Adelaide, 19 May 2023

About the Author

Nesimi Ertuğrul is an accomplished Electrical and Electronics Engineer, holding a BSc, Master's, and PhD in the field. He has been with the University of Adelaide, Australia, since 1994, where he is an A/Prof, specializing in a variety of applications of power electronics—an essential technology in renewable energy, battery storage systems, and electric vehicles. His research interests also extend to interactive computer-aided teaching and learning systems, remote access, and distance learning laboratories, involving object-oriented programming and signal conditioning, as well as data acquisition.

Dr. Ertuğrul has overseen the research of more than 30 PhD and Master's students and has been the chief investigator on numerous grants, funds, and consultancy projects, generating significant research income. Additionally, he has been the guest editor for three scientific journals.

With two solo-authored books, four book chapters, more than 195 journal and conference papers, and five patents to his name, Dr. Ertuğrul has made substantial contributions to his field. He has chaired international conferences, delivered keynote speeches, and served as a technical committee member. As a senior member of the IEEE, he has also served as an Associate Editor for the *IEEE Systems Journal*.

Dr. Ertuğrul has been featured in various media outlets, including radio and TV stations, newspapers, and social media platforms, primarily discussing the future of power systems, electric vehicles, battery storage systems, and renewable energy.

Outside of his professional achievements, Dr. Ertuğrul has a strong passion for wood-fired oven building and cooking, a hobby he enjoys sharing with friends and loved ones. Over the past 18 years, he has blended his engineering background with his love for cooking, incorporating advanced measurement and analysis techniques into his culinary endeavours.

"Aerial view images of excavations at Göbekli Tepe" D-DAI-IST-GT-2010-NB-5687.tif Deutsches Archäologisches Institut.

Photographer: Nico Becker.

History

1

1.1 INTRODUCTION

The wood-fired oven holds a prominent place in human culinary history, with its use tracing back to ancient times. Its basic structure and heating methods have remained largely unchanged throughout centuries of utilization.

Writing systems were developed independently at least four times in human history. The first known occurrence was in Mesopotamia, between 3400 and 3300 BC. Subsequent systems emerged in Egypt around 3200 BC, late Shang-Dynasty China in 1300 BC, and in Mesoamerican cultures between 900 and 600 BC. The existence of bakeries and bread ovens in these early societies is well-documented through artifacts and tablets from these eras [1].

Among the world's earliest urban cultures, which developed between the Tigris and Euphrates rivers, wood-fired ovens were a common culinary tool. Nevertheless, the first documented wood-fired oven structures can be traced back to the early Egyptian period, amid the nascent settlements and agricultural practices around the northern Nile valley. There is also evidence of their use during the Roman conquest of Egypt around 30 BC.

Historical depictions of a wood-fired oven were discovered on the monument of Vergilius Eurysaces, a well-known baker from ancient Rome. The monument features a series of carvings illustrating the bread-making process and technology employed in the bakery at the time (see Figure 1.1). These images portray the acceptance of grain deliveries by state officials, the grinding of flour with the aid of slaves and donkeys, and multiple sieving and quality-checking processes. One carving shows a horse tethered to a kneading machine, walking in a circle while slaves assist with any dough obstructions. Additional carvings depict eight slaves shaping bread on two tables under the watchful eye of a supervisor. The bread is then baked in an igloo-shaped, wood-fired oven, before it is loaded into baskets, weighed, and recorded by state officials for distribution. This tradition of communal flatbread, or yufka, cooking is still a revered practice in Türkiye, known for producing the best quality bread.

1.2 POTTERY AND COOKING UTENSILS

Several centuries before the urban revolution and the formation of early states, Lower Mesopotamia experienced a significant revolution in agricultural technology [3]. Archival records suggest that bread played a central role in the civilization of Uruk (an ancient Sumerian city around 4000 BC), humanity's

DOI: 10.1201/9781032640136-1

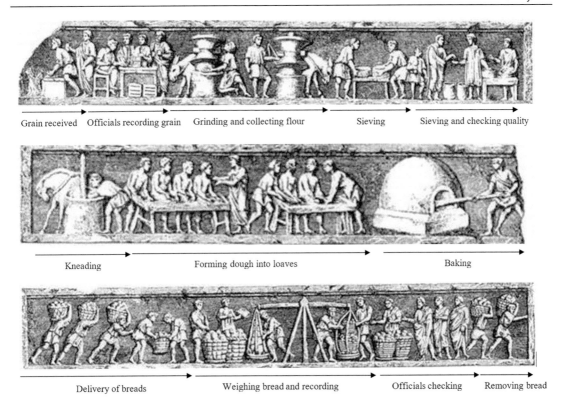

Grain received   Officials recording grain   Grinding and collecting flour   Sieving   Sieving and checking quality

Kneading   Forming dough into loaves   Baking

Delivery of breads   Weighing bread and recording   Officials checking   Removing bread

FIGURE 1.1   Roman time (100 BC) bread oven which is a part of a frieze of the monument to the baker Eurysaces [2].

From museum of old techniques at www.mot.be.

first complex society. An agricultural surplus, primarily from barley and wheat, in conjunction with this earliest urbanization, led to the invention of new tools and cooking techniques, including pottery-based utensils and ovens. Although pottery from Uruk was initially scarce, a variety of distinct ceramic derivatives later emerged in Southern Mesopotamia.

By 3000 BC, the Assyrians, natives of Mesopotamia, had introduced the art of pottery to the Hittites, who inhabited the ancient region of Anatolia (modern-day Türkiye). The same millennium saw the invention of the potter's wheel and the utilization of primitive kilns. During their settlements, the Hittites developed various cooking utensils and established an empire in Anatolia before 1700 BC. Hittite houses, built on foundations of fieldstones and upper courses of timber-reinforced unbaked mud bricks, contained indoor hearths or ovens for cooking [4].

Based on extant artifacts, pottery making in Anatolia commenced around 8000 BC near Çatalhöyük, a large Neolithic and Chalcolithic proto-city settlement in southern Anatolia, now recognized as a UNESCO World Heritage Site. Chalcolithic period pottery (4500–3500 BC) has also been discovered in nearby townships, including Hacıbektaş, Gelveri (Güzelyurt), and Avanos.

Still a centre of Anatolian pottery making, Avanos maintains its historical significance in Türkiye. The influence of various civilizations, including the Hittites, Phrygians (related to the Greeks), Romans, Byzantines, and Seljuks (a medieval Turkish Empire), can be traced in its pottery. The primary territory of the Hittite empire was bordered by the Kızılırmak River (also known as the Halys River and the Red River), Türkiye's longest river, which turns reddish due to the clay-laden volcanic soil from Avanos.

Clay and volcanic rocks, the primary materials in pottery and high-temperature bricks, can show a range of colours due to impurities. Notably, clay was also the material used to create the first known writing medium—tablets.

The entire region of Avanos, including the whole of Cappadocia, served as a refuge for early Christians fleeing Roman persecution, becoming a principal settlement centre. As pottery evolved, it produced various forms to support different cooking techniques. With its amalgamation of unique civilizations' cooking methods, this region likely holds the richest history of clay pot cooking techniques development in wood-fired ovens. Avanos continues to produce earthenware cooking utensils and pottery (see Figure 1.2), with a range of Avanos pottery-based cooking utensils to be illustrated later in this book.

Archaeological investigations have unearthed a diverse range of pottery-based kitchenware and structures associated with food preparation and storage. These include storage pithoi, cooking pots, tripod pots, stew-pots, kettles, baking plates, and cookers. Large baking plates, with diameters reaching up to 80 cm [5], have been discovered in various locations, including Ḫattušaš [6] and Kuşaklı [7]. Constructed from a blend of clay and limestone, these plates were fired to withstand fluctuating temperatures. It is likely that they were employed in bread-making and as bases for portable ovens.

Moreover, cylindrical cookers with open tops and bases have been found in Ḫattušaš [8]. These were likely used as stands for cooking pots on hearths.

FIGURE 1.2   A pottery maker with a foot-driven pottery wheel and a clay model of pottery maker in Avanos/ Türkiye.

Ertuğrul, Nesimi, Personal photograph, "Pottery of Pottery Maker, Cappadocia Region, Nevşehir Türkiye," May 2010.

1.3 COOKING STRUCTURES

Throughout human history, the concept of a fireplace has primarily served three functions: cooking, ritual activities, and heating. Different languages have their own words for fireplace, like "hearth" in English and "ocak" in Turkish, but they all carry similar meanings: a part of the floor where a fire is built, a home, a fireside, or a burning place.

A hearth refers to a fireplace commonly used for home heating and open-fire cooking, predominantly through radiated and convection heat. Typically built above the ground in circular, oval, U-shaped (as shown in Figure 1.3), square, or rectangular shapes, hearths are made of high thermal mass materials such as brick or stone, although mudbrick and clay are also used in regions where these materials are scarce. Cooking utensils like pots or thin convex or concave hot metal plates are placed on the hearth, with wood and dry dung (dry manure) serving as the primary fuel sources.

Circular, oval, and U-shaped hearths were prevalent during the Early Bronze Age (3000–2100 BC), as evidenced in sites like Demircihöyük and Karataş in Türkiye, and the Late Bronze Age (1200–1150 BC), as seen in İnandıktepe and Ortaköy [9]. Hearths were generally constructed in the centre of a common living space or along the wall, in the middle of houses, and even inside ritual buildings, such as in Kilise Tepe, Türkiye [9,10].

A tandoor, widely used for baking and cooking in Southern, Central, and Western Asia, and the South Caucasus, is a type of oven known by different names like tannour, Punjabi and Afghan tandoor, Armenian tonir, Azerbaijani and Turkish tandır, Türkmen tamdyr, and Uyghur tannur. This cylindrical, deep, bell-shaped oven has a narrow opening and is often made of clay (blended with straw, goat hair, or in the form of a large pot), brick, or even metal, and can be built either below or above ground level. In Eastern and Southeastern Türkiye, tandoors are typically built below ground level.

Tandoor structures have been discovered from the Early Bronze Age in Troy and the Late Bronze Age in Kültepe and Ortaköy, all in Türkiye [9,10]. They were fuelled by wood, coal, and any waste materials that could extend and control the heating process, including pottery shards.

FIGURE 1.3   On the left side, a hearth is showcased, utilizing a convex, thin hot plate for the preparation of gözleme, a savoury Turkish stuffed turnover often made with spinach or parsley combined with feta cheese. On the right side, a slow-cooked meat dish is being prepared in a tandir, an oven built below ground level, located in the basement of a restaurant in Istanbul.

Ertuğrul, Nesimi, Personal photographs, Istanbul and Mugla, Türkiye, 2010.

Tandoors serve both baking and cooking purposes, baking usually thin flatbreads affixed to the inner hot wall surfaces and cooking by placing a pot at the tandoor's base or hanging meat on hangers supported on the tandoor's edges, utilizing all three heat types: conduction, radiation, and convection. Tandoors have a chimney-like main assembly providing limited but effective ventilation.

The common fuel source for tandoors varied from wood to charcoal over time. A tandoor can reach temperatures as high as 480°C, requiring longer periods of lighting to sustain such elevated cooking temperatures. In below-ground-level tandoors, slow cooking is possible after the active fire has been extinguished and the oven opening tightly sealed (usually with mud, dough, or a damp piece of carpet). Notably, food in a tandoor oven may also experience some degree of smoking due to the fat and food juices that drip onto the heat source, exposing the food to live fire and radiated heat.

1.4 HISTORY OF COOKING

Though wood-fired ovens are often associated with baking bread and pizza, this book seeks to highlight the wide array of dishes that can be prepared in these ovens. The key lies in employing the correct oven utensils at appropriate temperatures and types of heat, enabling a multitude of cooking techniques beyond just bread and pizza. One such utensil is the clay stew pan (also known as clay pot or earthen pot), which draws upon the earliest form of cooking vessel, incorporating boiling and condensation into the cooking process.

It's crucial to note that, unlike modern stainless-steel pots, wood-fired ovens and clay pots naturally infuse food with earthy aromas and flavours, as well as the smoky taste and scent of the wood. These interactions between the food and the cooking vessel add an extra layer to the culinary process.

The culture of cooking and the development of recipes likely began with the invention of writing, as these aspects cannot be traced through archaeology or ethnoarchaeology. It's presumed that, parallel to the creation of pots for open fires, hearths, and wood-fired ovens, cooking methods also evolved. Some archaeologists propose that our prehistoric ancestors might have boiled food by dropping heated stones into the pot, or by cooking directly on rocks, as suggested by food remnants found in these stones [11]. Consequently, it's plausible that bread was also baked on stones in ancient times.

Although it's assumed that cooking techniques evolved independently in three primary regions of the world: Mesopotamia/Sumerian/Hittite, Mesoamerica, and Shang-Dynasty China (see Figure 1.4), future discoveries in Göbeklitepe may alter the timeline of cooking history in the coming decades. While there were some limited interactions between these early civilizations, pinpointing the origins of cooking and food recipes proves challenging due to the absence of written resources. Nonetheless, it's evident that today's diverse cuisines have emerged from these early practices and techniques. Wood-fired ovens, predominantly used for baking bread, have played a significant role in shaping the remarkable diversity of contemporary cooking styles.

The oldest known culinary texts, called the Yale Culinary Tablets, were found in Mesopotamia and date back to around 1700 BC. Comprising 35 recipes across three tablets, these are the oldest known recipes in the world [12,13]. A common characteristic across these recipes is the use of meat, fowl, vegetables, or grain, all cooked in water, which points to the significant role of boiling in Mesopotamian culinary practices.

Before the Yale Culinary Tablets were discovered, different cooking methods were employed, including using radiant heat in an oven, indirect heat in hot ashes, and direct flame exposure for broiling, grilling, or spit roasting. A fourth tablet from the Neo-Babylonian period was also found over a thousand years later.

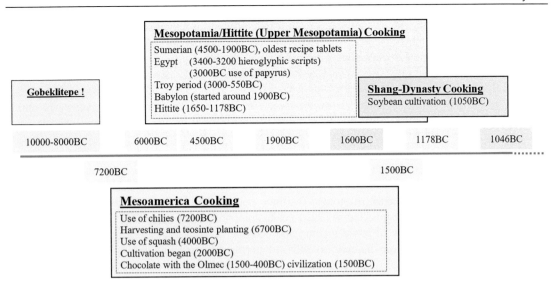

FIGURE 1.4 Cooking history and timeline.

Ertuğrul, Nesimi, Drawings.

The Sumerians, who embraced an agricultural lifestyle and techniques such as the plough, organized irrigation, large-scale intensive cultivation of land, and an organized agricultural labour force, utilized domesticated sheep, goats, cattle, and pigs in their cooking. Evidence for this is seen in primitive pictograms.

In Troy, or Ilium, an ancient city located in what is now Türkiye, many houses were fitted with a separate cooking area equipped with an indoor hearth or an oven-like structure. Excavations at the site uncovered two fire boxes, likely part of a baking oven. Some artifacts found at Troy, such as storage bins, contained remnants of carbonized wheat, suggesting wheat was a staple crop for the people living there. Milk and cheese were also commonly used in Troy's culinary practices.

Excavations at Hittite sites revealed that several varieties of wheat and barley were grown. Numerous Hittite properties also hosted a range of tree types, including vineyard vines, fig, pomegranate, almond, pistachio, and olive [14]. Sesame, linseed, and olive oil were likely used starting around 2500–2000 BC [15].

The Mesoamerican civilization, which developed in Mexico and Central America's regions, can be considered a New World equivalent to the civilizations of Mesopotamia and China. Mesoamerican cuisine has roots in these regions, where crops such as cacao, maize, beans, tomato, avocado, vanilla, squash, chili, turkey, and dogs were domesticated as far back as 7000 BC.

The introduction of these "New World foods" to Europe and Asia didn't occur until much later, around 1500 AD. This introduction of Mesoamerican ingredients to other parts of the world sparked a fusion of culinary traditions, notably in wood-fired oven cooking, which has since become a significant aspect of global cuisine.

China, the third main region, boasts a culinary history that predates 3000 BC, when agriculture first began in the region. Rice and millet were among the initial crops grown. Noodles, made from millet, were invented in 2000 BC, and fish farming was established by 1500 BC [16]. Although wheat and barley were introduced much later from Mesopotamia, during the Shang Dynasty period, cooking techniques that are still used today were developed. These include cutting food into small pieces, cleaning food and cooking vessels, maintaining balance among ingredients, and introducing sauces for meat and seafood, as well as pairing vegetables and fruits with main dishes.

The oldest known clay pot in Chinese history traces back to the Neolithic Age (10000–4500 BC). However, the ding, a large bronze pot with three or four feet, was introduced much later, around

1200–1100 BC. The ding was designed to create a space for a fire underneath and to increase heating efficiency by maximizing the vessel's surface area, making it ideal for cooking pork, beef, or lamb. Chopsticks also made their debut during the Shang Dynasty period [17].

Recent archaeological discoveries at Gobeklitepe (Potbelly Hill), one of the world's oldest sites, assumed to be temples, near Şanlıurfa in Southeastern Anatolia, Türkiye, have challenged mainstream theories on the development of cooking techniques. The site dates back to the Pre-Pottery Neolithic (9500–8000 BC), a period commencing at the end of the last Ice Age. New findings suggest that humans developed a taste for bread, beer, and other carbohydrates before they domesticated crops, as evidenced by the grinding of grains [18].

The processing of cereals and their role at Göbeklitepe have been the subject of in-depth study [19]. The site has yielded many tools related to food processing, including grinding slabs/bowls, handstones, pestles, and mortars. The absence of storage facilities implies that food production was primarily for immediate consumption, with possible seasonal peaks coinciding with large work-feasts [20].

While definitive evidence of bread-making and cooking activities at Gobeklitepe remains elusive, the findings suggest that some humans established permanent settlements long before they began farming. Future archaeological discoveries may provide more insight into bread-making and/or oven-based cooking activities, as the remaining 11 new hills (on a 100 km line) in the same region have yet to be excavated, potentially within the next decade.

One of the earliest cookbooks in recorded history is *Apicius*, named after the famous Roman merchant and epicure Marcus Gavius Apicius (who lived around 14–37 BC). Apicius (officially titled *De re coquinaria* or *The Art of Cooking*) was compiled around 400–500 AD and contains over 400 recipes [21]. This period marked distinctive features of Rome, as the empire had reached its pinnacle. Apicius was an avid food enthusiast, collecting recipes and even funding a school for teaching cookery and promoting culinary ideas [22].

Although discussions persist about the uniqueness of the culinary practices in Papua New Guinea, no historical record supports this claim. Traditional cooking methods in Papua New Guinea include roasting over an open flame, steaming in bamboo tubes, boiling in earthen pots, and baking with hot stones. It's evident that the country's culinary traditions were shaped by the indigenous Melanesian population.

1.5 HISTORY OF BREAD COOKING

While we cannot definitively identify the original creator of bread, bread and bread-like foods have played a significant role in the culinary history of many civilizations. Mesopotamian cultures, specifically those located between the Tigris and Euphrates rivers, the Old Kingdom of Egypt, ancient Greece and Rome, and many neighbouring countries, particularly in the Middle East and Türkiye, have all upheld robust traditions of bread-making [23].

In ancient Egypt, beer and bread formed a staple of the diet. Although numerous varieties of bread and cakes are mentioned in historical documents, their specific ingredients and characteristics remain unknown. Illustrations in the tomb of Ramesses III in the Valley of the Kings, dating back to the 20th Dynasty of Egypt, depict various forms of bread in the bakery, including loaves shaped like animals [24]. These illustrations also depict the process of making flour, with whole grains ground using flat stones.

A gypsum wall panel relief discovered in an Assyrian camp, created between 865 BC and 860 BC, offers another early record of food preparation and cooking. As shown in Figure 1.5, the circular shape on the left side of the figure represents a fortified camp with four sections. The bottom-right section of the circle portrays a cook amid other daily activities.

Over time, baking techniques have evolved significantly. During the earliest times, heavy pottery bread moulds, crafted from coarse mudware, were employed for baking. Subsequently, cylindrical cone-

FIGURE 1.5 An outline drawing represents a gypsum wall panel relief found in an Assyrian camp. This artifact was created during the rule of Ashurnasirpal II, the third king of the Neo-Assyrian Empire, between 884 and 859 BCE [25,26].

shaped moulds were used within square hearths. By the New Kingdom period (1570–1051 BC), a novel type of oven had emerged, and flat-disk shaped dough became common for baking.

The oven that came into use during the New Kingdom period (1570–1051 BC) differed greatly from earlier bread moulds. This oven was a large, open-topped clay cylinder encased in thick mud brick and mortar, akin to a tandoor-style oven. Likely, flat dough was slapped onto the pre-heated inner wall, baked, and then peeled off, similar to the current method used in tandoor cooking. While the exact influence of this oven on bread-shaping practices remains unclear, it is known that various shapes of bread, including those resembling fish and humans, as well as disks and fans, were produced during this period. The dough's texture varied as well, whether intentionally or accidentally due to the inclusion of coarse grains, somewhat akin to modern multigrain breads. These varied shapes and textures of bread, discovered in New Kingdom tombs, indicate the importance of bread in Egyptian life and culture.

In contrast to the Egyptians, the Hittites were recognized for their pragmatic outlook on life [27]. They considered a broad range of life aspects, including food, demography, housing and furniture, family dynamics, labour, wages and prices, medicine, entertainment, climate and geography, natural resources, resource exploitation, sanitation, personal hygiene, transportation, and social life [28].

One of the most significant Hittite civilization sites is found in Boğazkale, located in Türkiye's Çorum province. This was the Hittite capital known as Hattusha. Hittite tablets recovered from this site mention approximately 180 types of bread- and dough-based pastries, each cooked into unique shapes and using diverse ingredients. Interestingly, over 300 wheat and barley-based bread variations are recorded within the Mesopotamian civilization [12]. The Hittite tablets also provided instructions for maintaining hygiene within cooking spaces, implying that the Hittites were mindful not only of their food's taste and variety but also of the safety and cleanliness of their cooking environment.

Hittite breads were categorized and named according to a variety of characteristics, including their shape, size, weight, region of origin, ingredients used, and preparation method. Some of the general terms included thin-bread, thick-bread, and breads shaped like crescents, ears, teeth, grape bundles, humans, sheep, boats, and wheels. Other types included hot bread, harvest bread, sweet bread, honey bread, oil bread, spicy bread, pomegranatebread, wet bread, wheat bread, oat bread, breads with sesame seeds or green peas and beer, table bread, bread with and without yeast, fresh bread, stale bread, cucumber bread, white bread, soldier bread (tayin), and ritual bread [29].

In the Hittite language, the term "ninda" was used to refer to bread, but it could be applied to any food made with flour. Some specific Hittite bread names incorporated this term, such as *ninda.imza* (made without flour), *ninda.mulati* (made from barley), *ninda.gur.ra* (with cheese and fig), *ninda purpura* (small bread), *ninda.ku* (sweet bread), and *ninda.zu* (hard bread, also known as teeth bread) [12]. This broad application of the term may be why the wood-fired oven is often referred to as a "bread oven."

After threshing the grain with oxen, the Hittites utilized the straw fractions in various ways, including as animal fodder, as a tempering agent for clay bricks, and as fuel for cooking ovens [1].

As previously noted, flatbreads play a prominent role in the culinary tradition of wood-fired oven cooking that has its roots in the Hittite heritage. Despite their similar shapes and ingredients, flatbreads are given unique names around the world, often associated with specific geographical locations. Examples include chapati (South Asia and Africa), focaccia (Italy), frybread (Navajo, Native American), lavash (Türkiye and Armenia, with and without yeast), matzah (Jewish), naan (India, West/South Asia), roti (South Asia), tortilla (Central America and Spain), pizza (Italy), pita (Mediterranean), and pide (Türkiye). Interestingly, the last three popular types of flatbread—pita, pide, and pizza—share phonetic similarities.

Pide is a widely enjoyed Turkish flatbread that currently exists in two forms: plain (often finished with an egg or flour wash and sprinkled with white or black sesame seeds) and stuffed with an array of toppings, discussed in greater detail in Chapter 6. Pita, on the other hand, is a pouched, two-layered flatbread that can be used plain or for creating wraps with various local ingredients. The earliest known written records and recipes of pita-like flatbreads were baked in primitive forms of the tandoor or hearth, constituting a basic part of the diet. While there is no mention of these two flatbread terms in ancient texts or any medieval documents, they may have emerged around the same time as pizza, if not earlier.

The term "pizza" was first recorded in a Latin text from the central Italian town of Gaeta, then under the Byzantine Empire, in 997 AD. However, modern pizza seems to have evolved from similar flatbread dishes in Naples.

Türkiye prides itself on an impressive variety of flatbreads, which are still widely enjoyed and cover the majority of the territory once occupied by the ancient Hittite Empire. The diverse range of breads [30] includes pide (which can be enjoyed with or without toppings or fillings), meaty bread, corn bread, pita bread, bazlama bread (a thicker, pita-style bread that is regularly flipped to ensure even baking on both sides, usually cooked on a flat hot plate over a hearth or in a wood-fired oven), gozleme (a filled flatbread with various ingredients, primarily cheese and spinach, baked on the convex side of a sac griddle over a hearth), yufka (used in the preparation of numerous Turkish pastries as a sturdy wrap for fillings, and also in its yeast-free version as a long-lasting dry bread, produced in a communal baking activity in Anatolia), beze (grilled directly on an open fire using a meshed wire), lavash (smaller and thicker than yufka, thin and crispy, puffing up high as it cooks, either on a sac griddle or in an oven), and sourdough flatbread ("eksili ekmek") with yeast cooked on a sac griddle and sometimes placed on the opening of a tandoor. In addition, a variety of sourdough lavash on a sac griddle is available, such as Ankara Ebelemesi, which is served after brushing both sides with butter, and Ankara Cizlamasi, which is brushed with butter on both sides during cooking and is also known as butter-bread or "yagli ekmek."

The gozleme version of yufka bread, filled with minced meat, spinach, or cheese, is popular in the Safranbolu region of Türkiye and is called "sac bukmesi." Lavash bread is ideally cooked in a tandoor oven and comes in a variety of textures, created using fingertips or a bakery roller with spikes, and with different ingredients depending on the region. For instance, in the Sivas region, it is brushed with milk, yoghurt, and egg for a richer taste.

As part of the ongoing evolution of Turkish cuisine, most plain flatbreads are now served with hummus, tzatziki, and various sauces and mezes (a selection of appetizers common in West Asia, the Balkans, North Africa, and other regions). Additionally, pieces of plain bread often serve as cutlery, making them an "edible spoon."

During the Han and Jin Dynasties (265–420 CE), many products were introduced to the Central Plains of China from the western regions via the Northern Silk Road, including carrots, walnuts, garlic, flax seeds, and cucumbers. Among these, Nang Bing, a type of flatbread brought by people from Central Asia during the Tang Dynasty (618–907 CE), had a significant impact on gastronomy [9].

1.6 HISTORY OF OVEN TECHNIQUES AND COOKING STRUCTURES

The history of cooking methods is marked by continual evolution. The earliest forms of these methods include hearths, which were typically constructed of stones and clay, and were used for cooking food over an open flame. Ovens, often built into the ground or constructed with stones and clay, were also among the earliest cooking structures. Additionally, cooking pits, lined with stones and covered with dirt to retain heat, were used for roasting meat and vegetables. Some of these structures were designed to be movable, catering to the lifestyles of nomadic peoples. As the need for higher efficiency and better utilization of available fuels such as hay, dry dung, and wood increased, cooking structures such as the tandoor and wood-fired ovens were developed.

Archaeological excavations in Western Anatolia and the Aegean Islands have uncovered a variety of these cooking structures dating back to the Bronze Age [10]. For instance, hearths were usually constructed from stones and clay and used for cooking food over an open flame. Cooking pits, often lined with stones and covered with dirt to retain heat, were typically employed for roasting meat and vegetables. Some of these structures were even designed to be portable to accommodate the nomadic lifestyle of certain people.

In a bid to achieve higher efficiency and better utilization of fuels, the concept of a wood-fired oven was likely introduced. These ovens, owing to their greater thermal mass, could utilize all types of heat effectively.

Interestingly, the first illustration of an oven with a dome structure was discovered on the monument of the Roman baker Vergilius Eurysaces. However, numerous archaeological discoveries in Anatolia have also brought to light many structures that serve as the first tangible evidence of "wood-fired ovens." These ovens, found in various forms such as circular, oval, horseshoe, and square shapes, have been dated back to the Bronze Age (3300–1200 BC) and the early urban civilization era. Sites such as Karataş, Elaziğ, Mahmuthisar, Ortaköy, İnandıktepe, and Kinet Höyük in Türkiye have all provided evidence of these structures.

One intriguing oven structure discovered in Ortaköy was constructed with two ovens adhered to walls and crossed by a hearth between them. This oven, featuring an earthenware and chambered, thermally insulated structure, had an opening or door for accessing the cooking platform—possibly designed for baking [31].

In another archaeological site, Bogazkoy in Türkiye, a movable bell-shaped vessel was discovered from the Hittite era. Made of two pieces with a flat plate of clay and a small dome (diameter: 49 cm; height: 30 cm), this vessel was likely used for cooking and baking purposes [12] (Figure 1.6a).

Recent archaeological findings in Türkiye have uncovered a domed oven in the ancient city of Troy that dates back 3,700 years. This significant discovery provides substantial evidence of Anatolian culture during and subsequent to the Bronze Age [32]. Intriguingly, the oldest known domed ovens, also uncovered in Troy (Figure 1.6b), have been determined to be from around 2000 BC, which is approximately 300 years earlier than this latest discovery. Further explorations in this region have shown similarities in the structure and function of hearths and ovens from these periods. A detailed summary of these findings, including the unique characteristics of these cooking structures, is presented in Table 1.1 [10,12].

In analysing the results, several conclusions emerge. The prevalence of pithos in oven structures may be attributed to its local abundance. Moreover, the dominance of dome-shaped ovens confirms a long-standing preference for robust structures. Notably, a dome shape in masonry ovens (which include

(a)

(b)

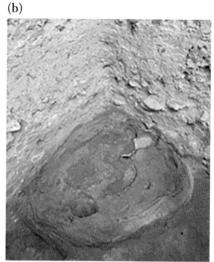

FIGURE 1.6   The remains of the best-kept Hittite oven found at Bogazkoy-Buyukkaya in Türkiye (a) [12] and the area of the domed oven in Troy (b) [32].
[12,32].

TABLE 1.1   The summary of cooking structures found in Western Anatolia and Aegean Islands around the bronze age (3300–1200 BC)

	HEARTHS	*OVENS*	*UNCLASSIFIED*
Circular shape	212	…	106
Oval shape	12	…	3
With corner shape	4	…	…
Dome with pithos	…	**28**	…
Clay dome	…	**157**	…

brick, stone, marble, granite, limestone, cast stone, concrete block, glass block, and adobe) is structurally optimal. This is due to a dome of equal thickness providing perfect compression without the tension or bending forces to which masonry is susceptible [33–39]. As detailed in Chapter 3, the dome structure in wood-fired ovens ensures relatively uniform and controlled heat distribution within the cooking space while also imbuing food with the natural flavours of burning wood.

Insights gleaned from archaeological sites have also provided valuable information regarding materials utilized in heat transfer, including sand, clay, ash, limestone, pebbles, and crushed ceramics or glass. These materials are still commonly used today due to their availability, affordability, and remarkable heat retention and insulation properties. For instance, ash barriers are extensively used to counter flame-radiated heat in wood-fired oven cooking. Fine sand serves as an excellent base for laying floor bricks without the use of mortar, also known as "floating." Lime, a durable additive used in mortar, aids in constructing the dome of a wood-fired oven that frequently expands and contracts. These materials have proven their durability over time and continue to be recommended for their effectiveness in heat transfer applications.

The earliest known civilization, the Sumerians (4500–1900 BC), recorded three distinct terms in their tablets for "ovens": dúruna, immindu, and ninindu. Interestingly, the term "oven" was first defined in Sumerian. It is also noteworthy that the terms "tandur," "tennur," and "fırın" (all referring to wood-fired ovens) originate in Farsi, Arabic, and Latin, respectively. The Akkadian term "tinūrum," used during the first ancient Mesopotamian empire between 2400 and 2200 BC, bears phonetic similarity to the Arabic "tennur."

In contemporary Türkiye, "tandir" is commonly used to describe a type of "fırın" (furnus) that is built below ground. However, there is evidence suggesting that wood-fired ovens and hearths served not only as cooking mediums, but also as integral to ritual practices, symbolizing the continuity of fire.

Domed-structured ovens, found across various archaeological sites, demonstrated longer heat retention periods, making them ideally suited for producing thicker and leavened bread. The design of a dome with a pithos opening could also be linked to the concept of a "tandir," as it too has a single access point. The "tandir" design can be viewed as a transitional form between a temporary earth oven and a horizontal-plane masonry wood-fired oven. On the whole, these varying oven and hearth forms have played a crucial role in the evolution of cooking techniques and culinary culture over time.

Archaeological records indicate a significant number of ovens built in the "tennur" style. Radiocarbon dating investigations have revealed that bread was baked both inside these ovens and over fire.

A variety of wood-fired ovens have seen use across the Fertile Crescent region and Central Asia. However, this book concentrates on the half-sphere, igloo-shaped, or dome-shaped wood-fired oven, recognized for its remarkable efficiency and harking back to the earliest designs of high-efficiency ovens. This type of oven is adept at harnessing all forms of heat and has even influenced the design of contemporary industrial ovens. It is recognized globally under various names, including bread oven, pizza oven, earthen oven, clay oven, stone oven, black oven, guvec oven (named after a Turkish casserole or stew dish prepared in an earthenware pot), and more recently, the Tuscan oven (featuring a significantly higher dome) and the Cart/Trailer/Catering oven (utilized for catering parties and outdoor festivals).

Figure 1.7 showcases a wood-fired oven located near the town of Assos, Türkiye, that boasts an original front structure and a refurbished dome constructed from local volcanic rocks. Interestingly, this oven has no documented history. It is important to note that the historical record of Assos stretches back to the Bronze Age, with some ancient sources suggesting the town was established by the Methymians from the Island of Lesbos in 700 BC. The town was built atop an andesite volcanic rock, a fine-grained igneous rock celebrated for its exceptional heat retention qualities. Later sections of this book will elaborate on how the heat retention capabilities of oven-building materials are crucial for creating a high-quality wood-fired oven possessing an ideal thermal mass.

In conclusion, the evolution of cooking platforms has corresponded with the diversification of cooking methods and ingredients worldwide. This has been influenced by a variety of cultural, geographical, and historical factors. Despite ongoing progress, certain trends can be identified, including the discovery of fire, the development of cooking tools, the agricultural revolution, global trade and commerce, and industrialization. It is challenging to categorize the world's regions based on the evolution of

FIGURE 1.7 One of the oldest functioning wood-fired ovens in Türkiye, near Assos.

Ertuğrul, Nesimi, Personal photograph, "The historical wood-fired oven, Assos, Türkiye," May 2010.

cooking due to the substantial variation and potential overlap in culinary traditions and techniques within and between regions. Intriguingly, the wood-fired oven style of cooking platform and its associated cuisines are recognized and utilized in virtually all significant culinary regions globally.

REFERENCES

[1] Ekine, N. G., Oykusu, T. E. & Unsal, A. *3uncu Basim*. Istanbul, Türkiye: Yapi Kredi Yayinlari, 2006.

[2] A history of the bread oven, available at https://www.mot.be/en/opzoeken/bakovens/geschiedenis/ geschiedenis-van-de-bakoven, accessed on 6/03/2023.

[3] Algaze, G. *The Uruk World System, the Dynamics of Expansion of Early Mesopotamian Civilization*. Chicago and London: The University of Chicago Press, 1993, ISBN: 0-226-01381-2.

[4] Hoffner, H. A. Daily Life among the Hittites, Edited by R. Averbeck, M. W. Chavalas, and D. B. Weisberg. *Life and Culture in the Ancient Near East*. Bethesda, MD: COL Press, 2003.

[5] Schoop, U. D. Hittite Pottery: A Summary, Edited by H. Genz and D. P. Mielke. *Insights into Hittite History and Archaeology*. Leuven: Colloquia Antiqua 2, 241–275, 2011.

[6] Schoop, U. D. Dating Hittites with Statistics. Ten Pottery Assemblages from Boğazköy-Ḫattuša, Edited by D. P. Mielke, U. D. Schoop, and J. Seeher. *Strukturierung und Datierung der hethitischen Archäologie: Voraussetzungen – Probleme – Neue Ansätze. Internationaler Workshop Istanbul*. İstanbul: BYZAS, 215–239, 2003.

[7] Mielke, D. P. *Die Keramik vom Westhang Kuşaklı-Sarissa 2*. Rhaden, 2006.

[8] Çorum archaeological museum, available at http://yesilirmakbasinmuseums.org.tr/corum/mn1.html.

[9] Golec-Islam, J. *The Food of Gods and Humans in the Hittite World*, BA Thesis. Poland: The University of Warsaw, the Faculty of History, the Institute of Archaeology, Index No: 300013, 2016.

[10] Uhri, A. *Bati Anadolu Erken Tunc Caginda Mutfak Kulturu Acisindan Ocaklar ve Firinlar*, Master Thesis. Izmir, Turkiye: Ege University, 2000.

[11] The Science of the Oven. *Herve*, Translated by J. Gladding, New York: Columbia University Press, 2009. Original Title: De la science aux fourneaux.

[12] Unal, A. *Anadolu'nun En Eski Yemekleri, Hititler ve Cagdas Toplumlarda Mutfak Kulturu*. Homer Kitabevi, 2007.

[13] The ancient Mesopotamian tablet as cookbook, available at https://www.laphamsquarterly.org/roundtable/ ancient-mesopotamian-tablet-cookbook, accessed on 27/03/20223.

[14] Macqtjee, J. G. *The HITTITS and Their Contemporary in Asia Minor*. Thames and Hudson, 1986.

[15] Unsal, A. & Pesinde, O. A. *Turkiyede Zeytin ve Zeytinyagi*. Yapi Kredi Yayinlari, 2019.

[16] Temple, R. *The Genious of China: 3000 Years of Science Discovery and Invention*. Published by Inner Traditions, 2007.

[17] Tripod cooking vessel (Li), available at https://www.metmuseum.org/art/collection/search/49897, accessed on 04/01/2022.

[18] Curry, A. *The Ancient Carb Revolution, Nature* (Vol. 594). Springer, 488–491, 24 June 2021.

[19] Dietrich, L. et al. Cereal processing at early Neolithic Göbekli Tepe, southeastern Turkiye. *PLoS One* 14(5), e0215214 (2019, May 1). 10.1371/journal.pone.0215214

[20] Dietrich, L. *Plant Food Processing Tools at Early Neolithic Göbekli Tepe, Archaeopress Archaeology*. Archaeopress Publishing Ltd, 2021. ISBN 978-1-80327-093-7 (e-Pdf).

[21] The Art of Cooking. *The First Modern Cookery Book, Composed by Maestro M. of Como*. University of California Press, 2005.

[22] Vehling, J. D. *Apicius, Cookery and Dining in Imperial Rome*. New York: Dover Publications Inc, 1977.

[23] Rubel, W. *A Bread: Global History*. Reaktion Books, 2011.

[24] Donald B. R. (Ed.). *The Oxford Encyclopaedia of Ancient Egypt* (Vol. 1, p. 197). Oxford University Press, 2001.

[25] Akyüz F. *Asur Ordusu (Yeni Asur Dönemi MÖ 911–612)*, PhD Thesis. Ankara: Gazi Üniversitesi Sosyal Bilimler Enstitüsü, 2018.

[26] Available at https://digitalcollections.nypl.org/collections/the-monuments-of-nineveh-from-drawings- made-on-the-spot-by-austen-henry-layard#/?tab=about

[27] Hoffner, H. *Daily Life and Culture in the Ancient Near East*. Edited by C. Averbeck and W. F. Bethesda. CLD Press, 95–118, 2003.

[28] Hoffner, H. A. & Hethaeorum, A. Food production in Hittite Asia minor. *Journal of Cuneiform Studies* 28(4), 243–246 (October, 1976). (4 pages), The University of Chicago Press.

[29] Hoffner, H. A., Alimenta Revisited, Akten des IV. *Internationalen Kongressefst ir Hethitologie Vtirzburg*, Herausgegebenv on Gernot \Tilhelm, 199–212, 4–8 Oktober 1999.

[30] Silk roads exchanges in Chinese gastronomy, available at https://en.unesco.org/silkroad/content/did-you-know-silk-roads-exchanges-chinese-gastronomy.

[31] The remains of a 3,700-year-old domed oven were discovered in the ancient city of Troy, available at https://arkeonews.net/remains-of-a-3700-year-old-domed-oven-were-discovered-in-the-ancient-city-of-troy/, accessed on 06/03/2023.

[32] Robison, E. C. Optics and mathematics in the domed churches of Guarino Guarini. *The Journal of the Society of Architectural Historians* 50(4), 384–401 (1991).

[33] Paulette, T. S. *Grain Storage and The Moral Economy in Mesopotamia (3000–2000 BC)*, PhD thesis. Chicago: The University of Chicago, 2015.

[34] Singer, C. J., Holmyard, E. J. & Hall, A. R. *A History of Technology* (Vol. I). Clarendon Press, Oxford Press, 1954.

[35] Rubel, W. *Bread: A Global History*. Reaktion Books, 2011.

[36] Metli, M. & Ulutürk, M. Ekmek Penceresinden Mezopotamya Medeniyetinin Kültürel ve Dini Hayatına Bakış. *Medeniyet ve Toplum Dergisi* 4(1), 76–99 (2020).

[37] Pollock, S. Politics of food in early Mesopotamian centralized societies. *ORIGINI* XXXIV, 153–168 (2012).

[38] Radner, K. & Robson, E. *The Oxford Handbook of Cuneiform Culture*. New York: Oxford University Press, 2011.

[39] Marks, G. *Encyclopedia of Jewish Food* (1st Ed.). Houghton Mifflin Harcourt/Wiley, 2010. ISBN 9780544186316.

Ertuğrul, Nesimi, Personal photograph, "The start of fire, Adelaide, Australia," September 2021.

Principles of Heat Transfer in Wood-Fired Ovens

2

2.1 INTRODUCTION

Fire is a natural event that transpires when a fuel source interacts chemically with the oxygen in the air, resulting in the production of light and heat energy. This energy is often manifested in the form of visible light. The hue of the flames in a fire can fluctuate based on the fuel's chemical composition and the fire's temperature. For instance, flames may appear blue when particular chemicals in the fuel release light, or they may range from yellow to red when light is emitted from the soot in the flame or the smouldering char.

In the context of a wood-fired oven, the heat produced by the ignited wood is disseminated to the surrounding surfaces through three separate mechanisms: radiation, convection, and conduction. Among these, radiation is the primary method. Cooking in a wood-fired oven involves employing one or a combination of these three heat transmission mechanisms, which can facilitate the cooking of a mixed substance either directly or indirectly (for example, inside an enclosure, a clay pot, or a tagine).

The heating duration of a wood-fired oven predominantly hinges on two aspects: the oven's thermal mass and the texture (such as the rate of humidity), type, quantity, and state of the wood utilized. Depending on these parameters, this heating process can span between 2 and 4 hours.

2.2 BURNING BEHAVIOUR OF WOOD

To employ wood as a primary fuel, it first needs to be ignited through external aids such as a lighter or a fire starter. Upon initial ignition, the porous wood material's self-heating can trigger open flaming. The fire-starting ease is influenced by the wood type, its thickness, and moisture content.

However, for open flaming to be initiated via the wood's self-heating, four conditions must be satisfied. Firstly, the wood should possess the capability for self-heating. Secondly, self-heating should be ample to induce thermal runaway. Thirdly, thermal runaway needs to initiate self-sustained smouldering, a slow-burning process generating smoke but lacking visible flames. Finally, the smoulder front should reach the wood's exterior [1]. In summary, wood can be ignited via self-heating, glowing (where a surface gets some heat, resulting in a sudden temperature rise), and smouldering combustion modes.

DOI: 10.1201/9781032640136-2

17

TABLE 2.1 Temperature ranges of wood pyrolysis and combustion

TEMPERATURE RANGE	DECOMPOSITION PROCESSES
>100°C	The evaporation of chemically unbound water.
160–200°C	Wood begins to decompose slowly, and non-combustible (as H_2O) gases formed
200–225°C	Wood pyrolysis is still very slow, the gases produced are mostly non-combustible
225–275°C	The main pyrolysis begins and flaming combustion occurs with a pilot flame.
280–500°C	Volatile gasses are produced (CO, methane, etc.) and smoke particles are visible, char forms rapidly as the physical structure of wood breaks down.
>500°C	Volatile production is complete and char continues to smoulder and oxidise to form CO, CO_2, and H_2O.

Ignition pertains to the commencement of combustion, which can manifest in different forms. Nevertheless, the primary outcome is that once ignition transpires, wood starts to undergo degradation through a process known as pyrolysis. This results in combustible gases that ignite to create a hot flame, often referred to as a "normal flame." In certain situations, the normal flame is preceded by a cool flame, which is barely discernible during daylight and has a temperature below 400°C.

When wood undergoes thermal decomposition or pyrolysis, it fragments into both inert and combustible gases [2]. To sustain a self-propagating reaction, the combustible gases must generate enough heat to continue producing volatiles, chemical components that vaporize easily. This process heavily depends on various interdependent chemical reactions, along with factors like charring and wood variability (including wood density and moisture content).

Owing to its low thermal conductivity and density, and its high specific heat, wood typically experiences multiple pyrolysis stages. This process can take place even without an oxidizer and can lead to a significant loss of mass, even at temperatures below 200°C.

The rigid char remaining post-pyrolysis impacts the heat transfer to the unburnt wood, and several of the char's physical properties also affect the ongoing thermal decomposition. Once porous char forms (visible as cracks), heat transfer mechanisms transition from conduction-dominated heat to radiative heat. The products of pyrolysis subsequently experience a swift combustion reaction known as ignition, which disseminates heat energy to the surroundings and can result in either smouldering or flaming combustion. Table 2.1 summarizes the vital temperature ranges for pyrolysis and combustion of wood [3].

During ignition, it's worth noting that the gaseous species can either be energized by a spark or flame (referred to as piloted ignition), or they can acquire the necessary energy for ignition solely through heating (known as unpiloted ignition). The ignition criteria are defined by either the "critical heat flux" (the minimum heat flux needed to trigger ignition) or the "critical surface temperature" (the lowest surface temperature at which ignition transpires). Table 2.2 offers typical values for these criteria, compiled from [2]. While temperature and radiative properties are substantial factors in the wood's combustion process, critical heat fluxes for piloted ignition generally fluctuate around 10–13 kW/m^2, whereas those for unpiloted ignition are around 25–33 kW/m^2.

Wood can ignite in two distinct modes: flaming and glowing. The ignition temperature for wood exposed to the minimum possible heat flux (approximately 4.3 kW/m^2) is around 250°C. A direct-flaming ignition transpires at surface temperatures between 300 and 365°C. However, wood can also ignite in a glowing mode, which may or may not transition into flaming. Presently, there is no comprehensive theory that includes the likelihood of glowing, glowing succeeded by flaming, or direct-flaming ignition modes. The majority of published studies have concentrated on radiant or combined radiant/convective heating.

TABLE 2.2   Some critical "heat fluxes" and "surface temperatures" of some typical woods

CRITICAL HEAT FLUX (KW/M^2)		CRITICAL SURFACE TEMPERATURE (°C)	
PILOTED	UNPILOTED	PILOTED	UNPILOTED
		382–405 (Mahogany)	
		372–395 (Red oak)	
			203–257 (sawdust)
10–12 (plywood)		296–330 (plywood)	
14.6 (Western red cedar)	26.8 (Western red cedar)		
14.6 (American whitewood)	25.5 (American whitewood)		
12.6 (African mahogany)	23.8 (African mahogany)		
15.1 (Oak)	27.6 (Oak)		
10.3 (Pacific maple)			
13.2 (Radiata pine)			
9.7 (Poplar)	28.90 (Poplar)	335 ± 59 (Poplar)	533 ± 63 (Poplar)

Another approach to understanding the ignition process in a wood-fired oven unfolds as follows: As the wood combusts, it liberates energy that elevates the temperature of the gases produced during combustion. Consequently, these gases radiate heat that warms up the interior bricks of the oven. The peak temperature of the flame is attained when two conditions converge: firstly, when all the energy from combustion is deployed to increase the temperature of the gases (a state referred to as "complete combustion"), and secondly, when the quantity of air supplied for combustion is ideally balanced.

In a wood-fired oven, air enters through the lower section of the door cavity and circulates within the oven due to convection. It then exits through the upper part of the door and enters the chimney (see Figure 2.1a). Preheated air present inside the oven further aids in augmenting the flame temperature.

The radiative properties of the combustion products in a wood-fired oven comprise a combination of gases and soot particles. The radiation emitted from luminous flames is the result of the collective contribution of these gases and soot particles. Non-luminous flames also originate from combustion gases, including carbon dioxide and water vapour.

Flames typically exhibit orange and yellow colours when burning traditional carbon-based organic fuel sources, such as wood, charcoal, and paper (see Figure 2.1b). However, it's important to note that not all carbon-containing fuels generate orange or yellow flames, and these flames generally have a temperature range of approximately 600–1200°C. The presence of these colours signifies partially burnt, or not fully ignited, carbon compounds.

In certain cooking techniques, including grilling and smoking meats, chefs often prefer orange and yellow flames because the smoke imbues the food with a savoury and flavourful taste. Bright blue flames, which usually range from 1260 to 1650°C, indicate the complete combustion of carbon, a phenomenon common in gas-burning fires.

It is worth noting that the presence of specific chemicals or compounds in woods—whether occurring naturally or added due to paint or wood treatments—can create different flame colours, such as green, pink, and red. Consequently, the colours orange, yellow, and blue represent a natural combustion process (complete or partial) of the fuel's carbon compounds, while other colours suggest harmful combustion and should be avoided for health and safety reasons.

Note that the ignition characteristics of wood are influenced by several factors, such as its moisture content, dimensions, and positioning during the combustion process. Moreover, when wood interacts with air to produce a flame, this flame tends to travel along the oven's inner dome surface, following the

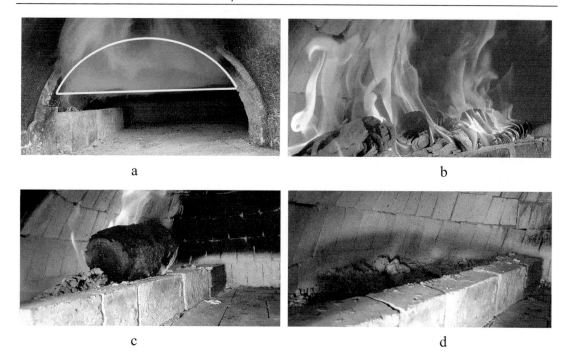

FIGURE 2.1 The air inlet region (the bottom section) and the smoky outlet region (the top section, high-lighted by the yellow border) of the door cavity during smouldering (a); orange and yellow flames are visible during wood combustion (b); reduction (gasification) (c); and at the end of the fire cycle, where ash is present (d).

Ertuğrul, Nesimi, Personal photograph, "The states of wood fire, Adelaide, Australia," December 2022.

path of least resistance and marking a region directly exposed to the flame. This region can be identified as the white portion of the fire bricks near the fire in Figure 2.1c.

While the processes of ignition, combustion, pyrolysis, and extinguishing are generally well-understood within controlled environments such as industrial ovens, predicting the burning behaviour and radiant energy transfer of wood within wood-fired ovens is a complex task. For this reason, an experimental method is used in this book to gain a deeper understanding of the impact of wood combustion on the heating characteristics of a specific wood-fired oven. This was achieved by employing thermocouples and a data logging system, details of which will be explored comprehensively in Chapter 5.

Gasification is a thermo-chemical process (shown in Figure 2.2) widely employed in the industry to transform biomass into a flammable gas known as producer gas or syngas. This gas comprises carbon monoxide, minor quantities of methane, hydrogen, water vapour, carbon dioxide, tar vapour, and ash particles. Although wood-fired ovens are not designed to operate as gasification plants, all four stages of gasification take place during the lifecycle of wood that is collected for use as firewood.

Before initiating slow cooking with the oven door closed, it might be necessary to extinguish an active fire with water to halt combustion and lower the oven temperature to a desirable range of 100–150°C. It's crucial to note that the slow cooking temperature in a wood-fired oven is almost identical to the "drying" temperature. Wood can be seasoned either outdoors or partially inside the oven to achieve this. If damp wood is used, it can produce volatile gases. Moreover, if the wood used contains volatile organic compounds such as paints, varnishes, wax, and solvents, it can release various chemicals that may pose health risks. Therefore, it's essential to avoid using wood containing these compounds.

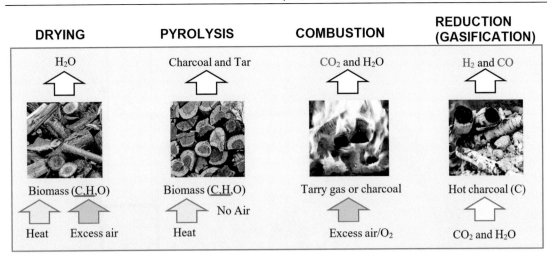

FIGURE 2.2 Four processes occur in thermal gasification, represented as conversion pathways.

Ertuğrul, Nesimi, Drawings with personal photos.

When initiating combustion, it's crucial to avoid stacking the wood too densely. If the bottom part of the pile does not receive enough air, pyrolysis can produce charcoal. Charcoal can also be produced when the fire is extinguished by dousing it with water for slow cooking before completing the "Reduction" stage.

It is worth noting that while the theoretical background of wood firing is provided here, all the information regarding the heating of a wood-fired oven should be seen as a flexible guideline or a starting point for a specific wood-fired oven cooking journey. Some of the tips provided may become more applicable and make more sense after conducting several firing and cooking practices in the oven, measuring temperatures and timing. This is ideally achieved by recording local temperatures and establishing the time constant of the oven structure.

2.3 RADIATION, CONVECTION, AND CONDUCTION

The movement of particles within an object, such as atoms and molecules, generates kinetic energy, which is recognized as heat or thermal energy. Temperature measures the average kinetic energy of these particles and serves to denote how hot or cold an object or its surrounding environment is. Although energy can be transferred without the temperature of a substance changing, generally, as heat energy increases, so does the temperature.

During the combustion process, radiation is the primary energy transport mechanism to the surrounding inner surfaces of the oven. This is followed by convection, which is heat transfer by the large-scale movement of fluid, and conduction, which is heat transfer through stationary matter via physical contact. Both conduction and convection necessitate matter for heat transfer. If there is a temperature difference between two systems, heat will transfer from the higher temperature system to the lower temperature system.

The three heat transfer mechanisms—conduction, convection, and radiation—all contribute to the heating of the wood-fired oven and the cooking of food within it. Figure 2.3 (top) demonstrates how these mechanisms operate in the oven. Conduction transfers heat through solid materials, while convection transfers heat via air. Radiation, on the other hand, transfers heat through electromagnetic waves

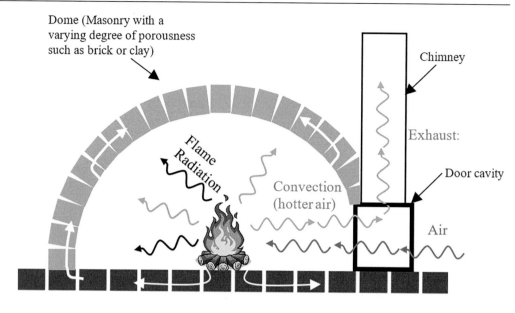

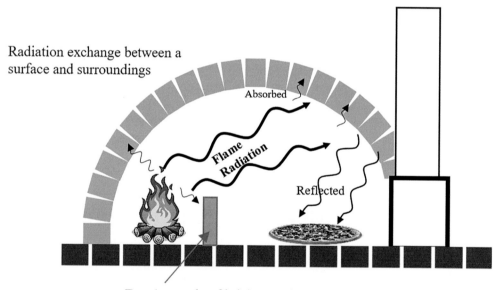

FIGURE 2.3   An illustration of the three distinct heat transfer methods that occur in wood-fired ovens during combustion (at the top), along with a representation of the distribution of the resulting radiated heat (at the bottom).

Ertuğrul, Nesimi, Drawings.

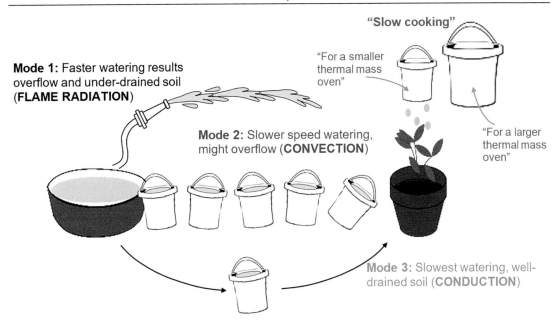

FIGURE 2.4   The analogy of watering a plant to understand the heat transfer phenomena in wood-fired ovens.

Ertuğrul, Nesimi, Drawings.

and can be felt as warmth near a fire. However, it can also be dangerous and cause serious injuries, so precautions must be taken to avoid direct exposure.

Figure 2.4 provides an analogy to better illustrate the different heat transfer rates. In this analogy, the plant symbolizes the food to be cooked, the water in the large bucket represents the heat source (burning wood), and the various watering methods represent the different modes of heat transfer. This analogy can also assist in understanding the relative efficiency of the different heat transfer mechanisms.

Each heat transfer mode can be briefly described as follows:

Mode 1: Heat transfer through radiation. In this mode, water is directed to the pot by a hose nozzle, independent of the medium. This method provides a quicker delivery of heat, but it can be uncontrolled and may damage the plant or cause an overflow.

Mode 2: Heat transfer through convection. In this mode, the water is transported to the pot through the medium by a brigade passing the water bucket. This method provides a slower delivery of heat compared to radiation.

Mode 3: Heat transfer through conduction. In this mode, the water is carried through the medium by a person running back and forth. This method provides the slowest delivery of heat among the three modes.

2.3.1 Slow Cooking

"Slow cooking" in the context of these analogies and heat transfer methods could be likened to the method that provides a gentle and sustained transfer of heat over a long period of time (see Figure 2.4, a single bucket dripping water). In this analogy, the small holes at the bottom represent the low-temperature setting commonly used in slow-cooking methods. Just as the small holes allow only a small amount of water to leak out at a time, the low temperature in slow cooking ensures that only a small amount of heat is transferred to the food at any given moment, allowing for gradual and thorough cooking. This can take several hours, but it is known to produce tender, flavourful dishes, particularly when using ingredients like tough cuts of meat.

The amount of water (or heat) in the bucket can be seen as an analogy for the thermal mass of the oven that is a material's ability to absorb and store heat. As it will be demonstrated later, the greater the thermal mass, the more heat the oven can absorb and store, and the longer it can continue to radiate that heat.

Note that when the door of a wood-fired oven is closed for slow cooking, it does alter the dynamics of heat transfer within the oven.

1. Convection: When the oven door is closed, the movement of air (and thus convective heat transfer) within the oven is somewhat limited compared to when the door is open. However, convection still plays a role in cooking as hot air circulates within the closed oven space.
2. Radiation: The oven walls, floor, and ceiling, having been heated by the fire, continue to radiate heat even after the fire has been reduced or extinguished and the door closed. This is because the fire bricks or stones have high thermal mass and can store and gradually release a lot of heat. This radiant heat is a major factor in the slow cooking process, effectively cooking the food over a prolonged period.
3. Conduction: This mode of heat transfer takes place when the food is in direct contact with the hot oven surface.

Note that, the aim of slow cooking is not to rush the process, but to allow flavours to develop over a prolonged period of gentle cooking. This method results in tender and flavourful dishes, particularly beneficial when cooking tougher cuts of meat. Similarly, in the analogy, the slow release of water ensures that the entire process is steady and controlled, matching the philosophy of slow cooking.

Furthermore, the "slow cooking" method has various effects on the food's chemical composition, which can impact health in several ways including preservation of nutrients; helping to break down fibres and proteins in food, making it easier to digest; reducing harmful compounds due to high-temperature frying or grilling; enhancing flavour; and a safer option since the risk of burns and fire is reduced.

Recognizing how heat transfer works in wood-fired ovens is crucial for achieving optimal cooking results and deciding the most suitable cooking methods, which will be explored in depth at the end of this chapter. The analogies shared above also offer insights into preferred cooking techniques. Therefore, it can be concluded that the "fastest" heat transfer method, flame-radiated heat, should typically be avoided during cooking. Regrettably, this kind of heat is quite common in many commercial and home wood-fired ovens, which can compromise the cooking quality of dough products.

2.3.2 Radiation (also known as heat, thermal or infrared radiation)

Thermal radiation, a form of heat transfer, can happen through any clear medium, be it solid like bricks, or fluid like air. It carries energy away from the source via photons in electromagnetic waves, which can be either absorbed, reflected, or transmitted. The emission of this radiation is in all directions, hence termed directional. Consequently, the radiated heat from a flame will hit other surfaces, like masonry bricks, and will be divided into portions of reflection, absorption, and transmission.

Figure 2.3 (bottom) showcases various thermal radiation transfer activities inside a wood-fired oven. The flame gives off radiated heat, which is partially reflected and absorbed at different areas of the oven due to radiation exchanges between surfaces and their environment. This radiated heat sparks convection and conduction heat transfers inside the dome structure and the floor material. These warmed surfaces also give off heat. However, it is worth noting that direct controlled cooking in wood-fired ovens should ideally avoid thermal radiation in flames.

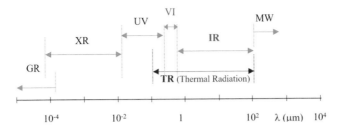

FIGURE 2.5   Electromagnetic spectrum classification according to radiation wavelength showing the wavelength range corresponding to thermal radiation. GR, gamma rays; XR, X-rays; UV, ultraviolet; VI, visible; IR, infrared; TR, thermal radiation; MW, microwaves.

Ertuğrul, Nesimi, Drawings.

The transfer of energy, or heat, can occur through three primary mechanisms: convection, conduction, and radiation [4]. Unlike convection and conduction, radiation does not require a medium to transfer energy. In fact, radiation is most efficient in a vacuum and is the fastest mode of heat transfer, moving at a speed close to that of light (approximately 3×10^8 m/sec).

Radiation proceeds along a straight line and is associated with both visible light and infrared heat, especially in the context of flames. Heat transfer through radiation occurs when electromagnetic radiation—in the form of infrared radiation (IR), visible light (VI), or even ultraviolet (UV)—is emitted or absorbed (see Figure 2.5) [4]. For instance, non-visible infrared radiation can carry heat directly from warm objects to cooler ones. This explains why we can sense the heat from a hot object even when we are not in direct contact with it.

A key example of radiant heat transfer is the Sun's warmth reaching the Earth, which is a mix of infrared and ultraviolet waves. Other examples include the heat released from a light bulb's filament or the flames in wood-fired ovens.

Flame heights are frequently used to estimate flame surface area and to measure radiation attenuation, which can be considerable over lengthy paths. However, in the dome structures of wood-fired ovens, this attenuation can be overlooked because the flame has a natural path that contributes to the heating mechanisms. As illustrated in Figures 2.1b and 2.1c, the intensity of radiation on the oven wall becomes substantial as the flame length increases [4].

In many combustion processes within wood-fired ovens, radiation is the dominant energy transport mechanism to the surrounding surfaces. Thermal radiation is deeply intertwined with the production of substances, such as soot, resulting from the incomplete combustion of organic matter, and the structure of the flame through local temperatures [5]. Consequently, characterizing radiative energy transfer is essential for modelling and understanding combustion processes, which tends to be more straightforward in industrial oven systems.

Comprehending the dynamics of radiative energy transfer in wood-fired ovens is vital for efficient and controlled cooking. Although characterizing radiative energy transfer tends to be simpler in industrial oven systems, it can pose challenges in traditional ovens due to their intricate geometries and diverse operating conditions.

Modelling the radiative properties of wood-fired ovens, indeed, presents a complex issue. The diverse structures of these ovens, in conjunction with variations in the type and quality of materials used, complicate the task of accurately determining the levels of absorption, emission, and scattering of radiant energy.

Given these challenges, this book does not seek to provide a theoretical approach grounded on a singular, comprehensive model to depict every wood-fired oven. Such an approach would be redundant, considering the unique features of each oven and the variability of the fuel (fire wood) used make it challenging to create a one-size-fits-all model. Furthermore, as it was highlighted earlier, radiative heat transfer may not always be the most preferred method of heat transfer for many cooking practices in wood-fired ovens.

2.3.3 Convection

Heat can be transferred from a surface via convection, which depends on several factors, including the surface temperature of the hot body, the ambient air temperature, the surface area, and the convection heat-transfer coefficient. In wood-fired ovens, convection currents facilitate heat transfer within the dome cavity and food container. The efficiency of convection is often quantified by the heat transfer coefficient (h), influenced by the thickness, type, and flow properties of fluids and the thermal conductivity of the medium through which heat is being transferred.

A larger heat transfer coefficient implies more efficient heat transfer from the source to the other medium. For instance, the convective heat transfer coefficient for air generally lies within the range of 10–100 W/m^2K, while for water, it spans between 500 and 10,000 W/m^2K, as outlined in Table 2.3. It is worth noting that the heat transfer coefficient is inversely proportional to the thermal R-value, a metric that gauges an insulator's resistance to heat flow and is widely used for building materials and clothing insulation.

Importantly, while the heat transfer coefficient (H) does depend on temperature, its temperature dependence tends to be relatively weak for convection heat transfer.

In a conventional kitchen oven without a fan, heat transfer predominantly transpires through radiation from the oven walls, with a minor contribution from natural convection due to temperature disparities. This mode of heat transfer mirrors what is seen in wood-fired ovens. In contrast, a convection oven, equipped with fans, enables more efficient heat transfer and circulation through convective heat transfer.

In wood-fired ovens with an open door, natural convection plays a crucial role. Hot air ascends, carrying heat away from the source and circulating within the oven dome before exiting through the door cavity and chimney. This natural convection is a vital factor for efficient oven heating as it facilitates even heat distribution throughout the oven cavity.

Convection heat transfer in wood-fired ovens can be categorized based on the nature of the airflow. This can include forced convection with gases or liquids, free convection with gases or liquids, and latent convection involving phase changes such as boiling or condensation, as demonstrated in Table 2.3.

TABLE 2.3 Types of convection heat transfers and typical heat transfer coefficients

CONVECTION PROCESS	HEAT TRANSFER COEFFICIENT, H (W/m^2K)	REMARKS
Forced convection Gasses Liquids	10–250 100–20,000	This heat flow is caused by atmospheric wind.
Free convection Gasses Liquids	2–25 50–1000	Airflow is prompted by density differences caused by temperature changes, such as those that occur when air comes into contact with food, a pot, or a container. As the temperature of the air increases, its density decreases, leading it to rise and be replaced by cooler, denser air. In a wood-fired oven, if the door is left open, both free convection and a degree of forced convection coexist. However, if the door is partially closed from the top, forced convection can be limited, a practice commonly employed in commercial ovens.
Hidden convection Boiling or condensation	2500–100,000	This type of convection is associated with a phase change between the liquid and vapour states of a fluid, such as during boiling and condensation.

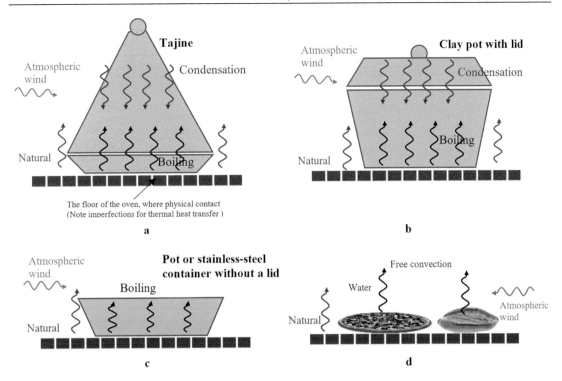

FIGURE 2.6   Illustration of types of convection heat transfers in the wood-fired oven cooking. a) A tajine pot over a heat source, where the heat causes the food inside to boil and the conical shape of the lid promotes the condensation of steam, which then drips back onto the food, creating a self-basting process. b) The clay pot with a flat lid allows for boiling and condensation. The steam rises, condenses on the underside of the lid, and returns to the food, but the flat shape may lead to less efficient self-basting compared to the conical tajine. c) An open cooking vessel without a lid, such as a pot or stainless-steel container, where the heat causes the liquid inside to boil, but without a lid, there is no condensation cycle to return moisture to the food. d) A pizza or a bread baking on a stone in an oven, where the heat source causes the bottom of the pizza to cook through direct contact with the hot stone (conduction), while the ambient heat of the oven cooks the top of the pizza by free convection. The atmospheric wind indicates the movement of air.

Ertuğrul, Nesimi, Drawings.

These three types of convection heat transfer can also be described concerning the airflow within the cooking items, as illustrated in Figure 2.6.

Latent convection occurs when warmer sections of a liquid or gas rise to cooler areas, and cooler liquid or gas replaces these warmer sections. Boiling water is an excellent example of such convection currents, represented in Figure 2.6 as boiling and condensation. Some cooking utensils, like Turkish Guvec, Moroccan Tajine, or German Romertopf, feature deep or shallow bases and domed lids. Latent convection takes place in these utensils due to fluid motion induced by vapour bubbles generated at the bottom of the boiling liquid-filled clay pot, as well as by the condensation of water vapour on the inner part of the domed lid. This is because phase changes of substances also induce convection currents.

Heat transfer coefficients can vary significantly under various convection processes. For instance, forced convection involving a low-speed flow of air over a surface may have a heat transfer coefficient of around 10, while a vertical plate in free air with a 30°C temperature difference may exhibit a heat transfer coefficient of 5. On the other hand, boiling water inside a container (latent convection) can have a heat transfer coefficient as high as 100,000 due to the significant heat transfer during phase changes.

Phase changes can be explained using a characteristic curve of temperature versus heat for a substance, such as water, as displayed in Figure 2.7. This curve is applicable to similar substances, though

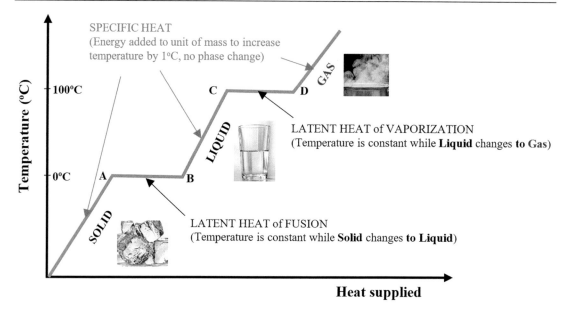

FIGURE 2.7   Phase changing characteristics of water.

Ertuğrul, Nesimi, Drawings.

with distinct slopes and critical points. It helps illuminate the phase changes that transpire when a substance is heated or cooled and how these changes can induce convection currents that enhance heat transfer.

Observe that when water in its solid form (ice) is exposed to heat, the temperature rises until a particular point, Point A. At this juncture, the temperature remains at 0°C despite increasing heat. The heat supplied from Point A to Point B is termed the latent heat of fusion. The energy provided from Point A to Point B serves to transition the solid (ice) into a liquid. The energy supplied beyond Point B is solely used to increase the liquid's temperature until Point C, where the water temperature reaches 100°C. Likewise, the quantity of heat provided from Point C to Point D is called the latent heat of vaporization, where the temperature stays constant while the liquid form changes to gas. Finally, if heating continues, all the energy beyond Point D is used to elevate the temperature of the gas. As mentioned, such phase changes occur fully or partially in substances used in cooking.

It is crucial to note that the time required to reach the critical points in the characteristic curve can vary depending on whether the heat is trapped inside a container, which reduces heat losses and nullifies atmospheric pressure, such as in boiling water when the lid is closed. With domed lids on clay pots, it is worth mentioning that they are usually unglazed. Due to latent convection, heat variation in the lids can be swift and high, so they must be soaked in water before usage. This pre-soaking enables the lids to absorb water and create a blanket of steam both inside and out, preventing cracking and aiding in cooking.

Moreover, using a clay pot with a lid enables greater heat transfer in all directions, as shown in Figures 2.6a and 2.6b. This results in quicker and more uniform cooking through phase change, increased heat, and evenly dispersed heat. In instances of direct cooking, such as baking bread, pizza, or lahmacun, no latent convection (boiling or condensation) occurs, only evaporation. In such scenarios, the moisture level, free and forced convection, and conduction heat transfer all contribute to the cooking time and the quality of the final product.

In general, convection heat transfer provides more uniformly distributed heat, and many temperature-sensitive foods like bread, vegetables, and clay pot cooking can greatly benefit from this method. However, some foods, such as soufflés, cakes, and muffins, may not be suitable for convection cooking, as the airflow can disrupt the delicate structure of these foods as they rise. Despite this, it is possible to cook such foods in a wood-fired oven using free convection with a closed door, which will be discussed further in Chapter 5.

As indicated in Table 2.3, slow cooking can be achieved by either halting the burning of wood or removing it completely, reducing the oven temperature to a desirable level, and then closing the door.

2.3.4 Conduction

Conduction heat transfer occurs between two solids that are in direct contact with each other. When particles in one solid gain more kinetic energy from radiated heat, absorbed heat, and convection heat, they transfer some of their energy to nearby particles in the second solid, with the process continuing. The flow of energy, or temperature rise, is from warmer areas to cooler areas.

In wood-fired ovens, conduction heat transfer takes place on the oven's floor, which is typically constructed from high-density fire bricks or high-temperature paving bricks. The efficiency of heat transfer depends on the quality of the physical contact surface between the floor bricks and the cooking utensils or bakery items. It should be noted that in some cooking practices, this contact surface may be deliberately reduced to either avoid overcooking or achieve a desirable texture, such as when cooking flatbreads on pebbles.

2.4 HEAT TRANSFER, SPECIFIC HEAT CAPACITY, AND EMISSIVITY

In a wood-fired oven, heat transfer depends on three primary factors: the temperature change, the thermal mass of the oven system, and the properties of the food being cooked, including its phase. Figure 2.8

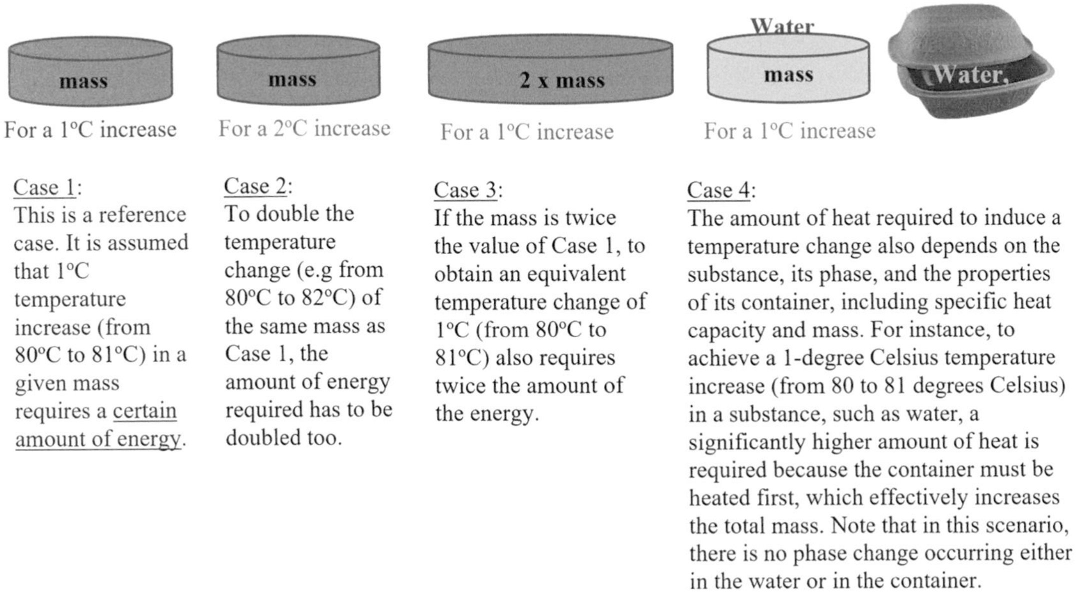

mass

For a 1°C increase

mass

For a 2°C increase

2 x mass

For a 1°C increase

Water
mass

For a 1°C increase

Water

Case 1:
This is a reference case. It is assumed that 1°C temperature increase (from 80°C to 81°C) in a given mass requires a <u>certain amount of energy</u>.

Case 2:
To double the temperature change (e.g from 80°C to 82°C) of the same mass as Case 1, the amount of energy required has to be doubled too.

Case 3:
If the mass is twice the value of Case 1, to obtain an equivalent temperature change of 1°C (from 80°C to 81°C) also requires twice the amount of the energy.

Case 4:
The amount of heat required to induce a temperature change also depends on the substance, its phase, and the properties of its container, including specific heat capacity and mass. For instance, to achieve a 1-degree Celsius temperature increase (from 80 to 81 degrees Celsius) in a substance, such as water, a significantly higher amount of heat is required because the container must be heated first, which effectively increases the total mass. Note that in this scenario, there is no phase change occurring either in the water or in the container.

FIGURE 2.8 The concept of heat transfer under four distinct cases.

Ertuğrul, Nesimi, Drawings.

TABLE 2.4   Specific heat capacities of some selected substances

SUBSTANCE	SPECIFIC HEAT CAPACITY (KJ/KGK)	SUBSTANCE	SPECIFIC HEAT CAPACITY (KJ/KGK)
Air (room condition)	1.01	Iron	0.41
Water	4.18	Clay pots	0.87
Water (ice)	2.05	Glass pot	0.84
Vegetable oil	2.0	Aluminium pot	0.9
Sodium (solid)	1.23	Copper pot	0.38
Beef (ramp-hamburger)	2.6–3.5	Cast Iron pot	0.46
Chicken	2.72–3.35	Fire bricks	1.05
Lamb (rib cut-carcass)	2.55–3.06	Granite	0.79
Bacon	2.01		
Fish	3.6		
Onion (dry-fresh)	0.87–4.01		
Kidney bean (dry)	1.17		
Potato	3.43		
Eggplant	3.94		

illustrates the concept of heat transfer under three distinct scenarios to accomplish a 1°C temperature change in the food being cooked. These scenarios include a uniform solid mass (Case 1), a solid mass twice as large (Case 2), and a different mass of a substance (water) inside a clay pot.

The quantity of heat transferred is proportional to the mass of the food and the temperature change. Furthermore, the specific food being cooked also influences the amount of heat transferred. For instance, the heat required to raise the temperature of alcohol is less than that needed for water. The phase of the food—whether it is a gas, liquid, or solid—also affects the quantity of heat transferred.

The term "heat capacity" is commonly used to quantify the heat transfer characteristics of a solid or gaseous substance. It represents the amount of heat needed to increase the temperature of the substance by 1°C. However, since this quantity is mass- or volume-dependent, the term "specific heat capacity" is more often used. This term describes the amount of heat needed to increase the temperature of a unit mass of a given substance by 1°C. The SI units for specific heat capacity are Joules per kilogram per degree Celsius (J/kg°C), which is equivalent to Joules per kilogram per Kelvin (J/kgK) or Joules per gram per degree Celsius (J/g°C).

It's important to note that the freezing and boiling points of water have a 100-degree difference on both the Celsius and Kelvin scales, meaning they have the same magnitudes. Table 2.4 provides the specific heat capacities of various substances, including those related to the structure of wood-fired ovens and basic cooking ingredients.

Note also that these definitions and quantities are given for homogeneous solid and gas substances only. In any cooking practice using a wood-fired oven, it is impossible to precisely define the specific heat capacity of a mixed set of ingredients (which contain various chemical compounds in ever-changing ratios) and diverse cooking utensils. Furthermore, the uncertainty in quantifying the heat generated from burning wood, coupled with the mixture of various ingredients, complicates the determination of the precise amount of heat transfer during cooking. For instance, raising the temperature of 1 kg of water by 1°C requires 4.18 kJ of energy, while vegetable oil requires 2.0 J of energy, which is less than half the amount required for water.

This difference is due to the distinct types of molecules that make up water and oil, and the energy required to increase their kinetic energy and change temperature. For example, the specific heat capacity

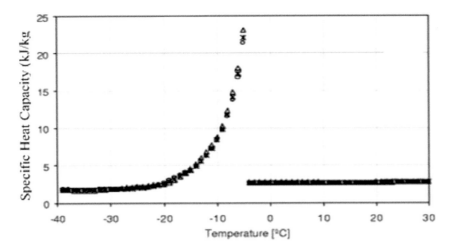

FIGURE 2.9   Experimental specific heat capacity of bread dough with different water content [6].

Ertuğrul, Nesimi, Drawings.

of a clay pot, which is 0.87, compared to that of a cast iron pot, which is 0.46—about half that of the clay pot. This implies that the clay pot can store up to two times more heat than the cast iron pot and has the capacity to release it slowly to the food inside.

It is also worth noting that the water content and ingredients of each dough formulation can vary significantly during cooking. Experimental data in Figure 2.9 underscores the significance of temperature when the dough is in a frozen state, as seen in ready-made frozen pizza. This highlights the importance of heat transfer and specific heat capacity of a formulated substance as the temperature of the substance, as well as the temperature of the wood-fired oven, changes during the cooking process.

Achieving the same texture in every wood-fired oven cooking, even with an identical formulation, is impossible, unlike in commercial ovens. Therefore, the specific heat capacity ratios provided in Table 2.4 should be used as a rough guide for comparison purposes, specifically in the context of wood-fired oven cooking.

Surface emissive power refers to the rate at which energy is radiated per unit area. Gaseous combustion products in flames emit radiation in discrete spectral bands [7]. Emissivity of a surface quantifies the amount of infrared energy radiated from an object. Flame emissivity, used to characterize an open pool fire, can be measured through two methodologies: by calculating the mass burning rate or by using infrared thermography, which captures and analyzes thermal information from non-contact thermal imaging devices.

Emissivity is essentially a measure of a material's ability to emit thermal radiation, often referred to as radiating efficiency. It is used as a modifying factor in thermal imaging cameras to attain accurate temperature readings, which will be explained in more detail in the Data Logging Chapter. It's important to note that emissivity and porosity are correlated, with porous materials typically exhibiting higher emissivity values. For instance, earthenware, a porous material, has an emissivity value of 0.9, whereas commercial copper, which is less porous, has an emissivity of 0.07.

2.5 THERMAL MASS AND TIME CONSTANT

As shown in Figure 2.3, radiation can occur on a surface from its surroundings. In a wood-fired oven, radiation primarily originates from the wood source, predominantly in the form of flame radiation.

A portion of this radiation is absorbed by surrounding surfaces, thereby increasing the thermal energy of the surface material, while a part of it is reflected, contingent on the nature of the radiation and the surface material. To put it differently, the radiation emitted by the walls of the wood-fired oven is largely dependent on their absorptivity.

As a user of the wood-fired oven, the primary goal is to heat the oven to a desired temperature with the maximum amount of heat being transferred to the food being cooked, and as little as possible escaping with the flue gas, all while minimizing the production of NOx. Moreover, the key aim is to maintain a stable temperature region in the wood-fired oven for a sufficient duration to achieve predictable textures in the cooked items. The definition of thermal mass can be applied in the design of a wood-fired oven to illustrate how the mass of the entire oven structure provides inertia against fluctuations in internal temperature and heat losses.

The construction of the oven should consider both the building method and the exposure of the final dome structure to external environmental conditions. Consequently, commercial wood-fired ovens are not directly exposed to the environment. Additionally, the materials chosen and how they are used play a vital role in determining the thermal (inertial) mass of the final oven, which encompasses the dome and floor material, binding material, mortar, and insulating material (to be discussed in more depth in the next chapter).

It is important to note that in a commercial oven structure, accurate modelling and calculation of the thermal mass is feasible due to the uniformity of oven material and construction practices, along with the precision and predictability of heat transfer resulting from a uniform and predictable heat source like gas.

However, modelling a wood-fired oven is a highly complex task due to the diverse range of oven structures and materials used, along with the intricate heat sources (wood) and heat transfer mechanisms discussed earlier. Generally, thermal inertia can be described as the degree of delay with which the body temperature approaches the ambient temperature or the resistance of the material to changes in temperature. This results in transmission losses in the absence of an active heat source, as seen in slow cooking in a wood-fired oven.

Thus, a simplified first-order model of a wood-fired oven could be viewed as a heat storage system, where all materials contribute to the storage of heat. In this heat storage structure, there are transmission losses through the dome and floor, along with ventilation losses through the door cavity. Consequently, the term thermal inertia is occasionally referred to as a heating rate imbalance. With data logging systems now available, temperature changes can be recorded at second intervals and used to determine the "time constant," as quantified using measured data in Chapter 5.

The time constant can be influenced by the thermal masses that store heat, transmission loss, and ventilation loss. Enhancing the heat storage capacity of a wood-fired oven implies extending the duration of heat retention, or increasing the thermal time constant. Therefore, increasing the thermal time constant (and thus the thermal mass) in wood-fired ovens can be achieved by reducing heat losses. This underscores the importance of insulation, in addition to the heat retention capability of multiple structural materials. Figure 2.10 illustrates two sets of sample temperature data captured from a wood-fired oven with no active fire inside the oven and the door sealed for slow cooking practice. It should be noted that the rate of temperature change ($\Delta T/\Delta t$) is proportional to the time constant of the major oven sections measured: the dome and floor, which form the oven's thermal mass. It's also worth noting that, even with the door closed, the rate of temperature change differs because the oven sections possess different thermal masses and are exposed to the ambient environment in varying ways (see Chapter 5). However, when the door is open, the rate of temperature change in both sections accelerates due to substantial heat losses and external airflow into the oven cavity.

For instance, if the dome of a wood-fired oven is composed of two materials with contrasting thermal properties, they can be conceptualized as an insulating material (with high resistance to heat transfer) and a storage material (like brick). In this model, the insulation is considered purely resistant, linked to the surface heat transfer resistances. The brick layer is considered a pure capacitor (storage element). By employing a similar construction methodology to the floor, a lumped-heat-capacity method can be developed for a given wood-fired oven.

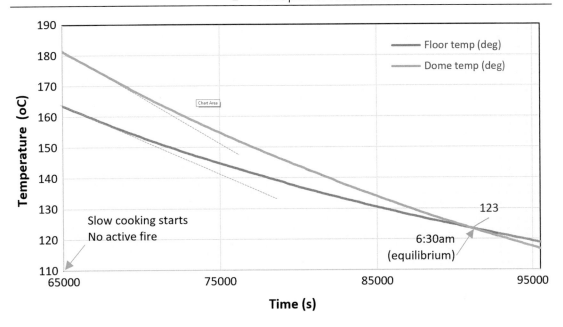

FIGURE 2.10 The rate of change of temperature (ΔT/Δt) in a wood-fired oven. On the dome: 0.0026°C/second (9.36°C/hour); on the floor: 0.0018°C/second (6.45°C/hour).

Ertuğrul, Nesimi, Drawings.

It's essential to note that, similar to building construction, the heat storage capacity of a wood-fired oven structure is increased if the storing mass is positioned inside the insulation [8,9]. This approach will be further discussed in Chapter 3.

2.6 COOKING METHODS AND HEATING MECHANISMS

This book explores various cooking techniques in the context of the heat transfer mechanisms elucidated in previous chapters. Direct heat cooking involves keeping a burning fire within the oven chamber, as seen in roasting or pizza cooking. On the other hand, retained heat cooking involves removing the fire prior to cooking, as demonstrated in slow-cooking meats or baking bread.

Table 2.5 illustrates a heuristic relationship between key cooking techniques and types of heat transfer, specifically relevant to wood-fired oven cooking. It is important to underscore that unique cooking methods can be crafted by utilizing different cooking utensils and a mix of these methods, facilitating the creation of distinctive flavours and textures. These methods will be explored in greater detail in Chapters 6 and 7, with numerical temperature ranges provided in Chapter 5, referencing the wood-fired oven under examination. The cooking methods outlined in Table 2.5 can be segmented into three main categories: dry heat cooking (baking, broiling, grilling, roasting, and sautéing), moist heat cooking (poaching, simmering, boiling, steaming, and frying), and combination cooking (braising, stewing, and searing).

Figure 2.11 summarises a qualitative comparison between temperature ranges and heat transfer mechanisms for various cooking techniques. Generally, higher heat transfer correlates with shorter cooking durations. However, given that the temperatures represented in the figure are qualitative

TABLE 2.5   Heuristic relationship between cooking methods and heat transfer types in wood-fired ovens

COOKING METHODS	CONDUCTION	CONVECTION	RADIATION
Baking (150–230°C) It is a cooking technique that primarily utilizes dry heat within the enclosed space of the oven, typically on the floor, or in hot ashes or on preheated hot stones. The application of heat to the food causes it to rise, brown, and cook evenly, making it particularly suitable for bread, pastries, cakes, and other baked goods.	Medium/Low	High/Medium/Low	Low
Broiling (230–300°C) It is a cooking technique that employs direct, high heat from a heat source located above the food. In this method, the food is placed on a rack or a broiler pan and cooked by being exposed to direct radiant heat, typically one side at a time, until it is browned and cooked through. This technique is frequently employed for cooking meats, fish, and vegetables, offering a rapid cooking process that achieves a crispy exterior while maintaining the tenderness and moisture of the interior.	Medium	Medium	Medium/High
Grilling (200–230°C) This is a cooking technique that typically relies on a substantial amount of thermal radiation. It involves using a grill, grill pan, or griddle to cook meat, vegetables, and fruits quickly and efficiently. The direct application of high heat imparts a distinctive flavour and char to the food, making grilling a popular choice for outdoor cooking and barbecues.	Medium	Medium	Medium/High
Roasting (160–230°C) Roasting is a dry heat cooking method using an oven, ideal for larger cuts of meat, poultry, and vegetables. It achieves a crispy exterior and a tender, juicy interior. With indirect heat from all sides of the oven, it's perfect for slow-cooking tougher cuts like goat at low temperatures (100–150°C) for hours, even up to 12. Roasting benefits from flame-radiated heat and convection, creating a crispy and dry surface on meats through direct heat transfer and circulating hot air.	Medium/High	Medium/Low	Low
Sautéing (180–200°C) Sautéing is a cooking technique that involves quickly cooking food in a small amount of oil or fat over high heat, which preserves its natural flavours and moisture. It is commonly used for small pieces of food, such as diced vegetables or thinly sliced meats, and produces a flavourful and evenly cooked outcome. The intensity of the flame and the type of pan used are critical for successful sautéing.	High	High	Medium

TABLE 2.5 (Continued)   Heuristic relationship between cooking methods and heat transfer types in wood-fired ovens

COOKING METHODS	CONDUCTION	CONVECTION	RADIATION
Poaching (70–80°C) It is a gentle cooking technique where food is cooked in liquid, usually pre-heated water or broth, at a low temperature just below boiling. It can be done on the cooktop or in the oven. This method is commonly used for delicate foods like fish, eggs, and fruits, resulting in tender and moist results. To poach, the food is placed in a pot or pan with enough liquid to cover it, and no additional fats or oil are required.	Low	Low	Low
Simmering (80–90°C) Cooking involves simmering food in a liquid kept just below the boiling point, with gentle, small bubbles. This is achieved by initially bringing the liquid to a boil and then reducing the heat, such as by placing the cooking utensil in a low-heat position, like near the door cavity. This method is commonly used for dishes like soups, stews, sauces, beef stews, and vegetarian curries that benefit from long cooking times to develop flavourful broths and tenderize tough cuts of meat.	Medium	Medium	Low
Boiling (Around 100°C) Food is cooked by immersing it in boiling liquid, such as stock, which releases steam. Altitude and water-soluble substances like salt and sugar can impact the boiling temperature. This method applies to all pot cooking with a covered lid. Heat transfer from the boiling liquid denatures proteins, breaks down starches, and melts fats, resulting in changes to texture, flavour, and nutrition. Boiling is commonly used for vegetables, pasta, grains, and soups.	High	High/Medium	Medium/High
Steaming (100–120°C) It is a cooking method that involves exposing food to steam. This is achieved by placing the food in a steamer basket or on a rack inside a covered pot or steamer, which can be made of stainless steel, glass, or cast iron. The steaming process does not require any added fat, making it a healthy cooking technique. It is a versatile method suitable for various foods, such as vegetables, fish, poultry, and dumplings.	High/Medium	High/Medium	High/Medium
Frying (170–190°C) In the oven, this cooking technique involves using a minimal amount of oil to coat the pan, resulting in a crispy food surface. It is commonly employed as a preliminary step for combination cooking. This method is suitable for preparing a wide range of foods, including meats, fish, vegetables, and even desserts.	High	High/Medium	Medium/High

(Continued)

TABLE 2.5 (Continued)   Heuristic relationship between cooking methods and heat transfer types in wood-fired ovens

COOKING METHODS	CONDUCTION	CONVECTION	RADIATION
Braising (100–170°C) Braising is a slow cooking method where food is cooked in a covered pot with a small amount of liquid like broth, stock, or wine. This moist-heat technique is especially suitable for tougher cuts of meat like goat, beef brisket, or lamb shanks, which become tender and flavourful through long hours of slow cooking in the braising liquid. The braising process often involves browning the meat to develop a crust and enhance the taste. Vegetables and aromatics are then added to the pot along with the braising liquid. Once the pot is simmering, it requires minimal attention, making it a low-maintenance cooking method, particularly suitable for wood-fired oven cooking.	Medium/Low	Medium/Low	Medium/Low
Stewing (70–80°C) This method involves slow-cooking food in a covered pot with a small or larger amount of liquid, such as water, broth, stock, or wine. Once the pot is simmering, it requires minimal attention, making it a low-maintenance cooking technique. It is commonly used for tougher cuts of meat or poultry, yielding flavourful gravies, as well as for fibrous vegetables, grains, and legumes. The slow cooking process allows the flavours of the ingredients to meld and intensify, resulting in a rich, complex taste and tender, juicy meat.	Low	Low	Low
Searing (200–230°C) Searing is a cooking technique used to quickly cook the surface of food at high temperatures, typically at the start of the cooking process. It is commonly employed with meats like beef, chicken, and fish. To sear meat, it is first patted dry to remove excess moisture, then placed in a hot pan, griddle, or grill to develop a visually appealing crust, as well as enhance flavour and texture. After searing, the meat can be further cooked using methods such as roasting, braising, or grilling. It is worth noting that searing can also be applied to other foods besides meat, such as vegetables, tofu, cheese, or even fruit, to add colour and flavour to the dish.	Medium	Medium	Medium

illustrations, most cooking methods cover a wide temperature spectrum. Apart from the actual heat transfer options, the placement of food within the wood-fired oven can be adjusted to control the heat experienced by the item being cooked. As it was stated previously, it is essential to note that due to the rapid nature of the uncontrolled heat transfer mechanism, flame radiation should be avoided in

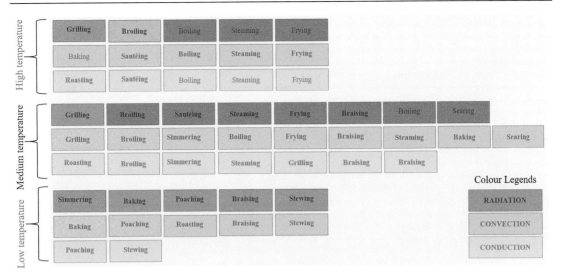

FIGURE 2.11 A qualitative comparison is drawn between temperature ranges and three heat transfer methods in conjunction with proposed cooking methods in the wood-fired oven.

Ertuğrul, Nesimi, Drawings.

wood-fired oven cooking. Lastly, it is noteworthy that, similar to microwave cooking, flame radiation can infuse heat into even very hot foods.

REFERENCES

[1] Babrauskas, V. *Ignition Handbook: Principles and Applications to Fire Safety Engineering, Fire Investigation, Risk Management and Forensic Science*. Fire Science Publishers, 2003.

[2] Bartlett, A. I., Hadden, R. M., & Bisby, L. A. A review of factors affecting the burning behaviour of wood for application to tall timber construction. *Fire Technology* 55, 1–49 (2019). 10.1007/s10694-018-0787-y

[3] Lowden, L. A. & Hull, T. R. Flammability behaviour of wood and a review of the methods for its reduction. *Fire Science Review* 2, 4 (2013). 10.1186/2193-0414-2-4.

[4] Tien, C. L. & Lee, S. C. Flame radiation. *Progress in Energy and Combustion Science* 8(1), 41–59 (1982).

[5] Sen, S. & Puri, I. K. *Thermal Radiation Modelling in Flames and Fires, WIT Transactions on State of the Art in Science and Engineering* (Vol. 31). WIT Press, 2008. 10.2495/978-1-84564-160-3/08.

[6] Matuda, T. G., Pessôa Filho, P. A., & Tadini, C. C. Enthalpy and heat capacity of bread dough at freezing and refrigeration temperatures. 2006 CIGR Section VI International Symposium on Future of Food Engineering, Poland, 26–28 April 2006.

[7] Incropera, F. P., DeWitt, D. P., Bergman, T. L., & Lavine, A. S. *Fundamentals of Heat and Mass Transfer* (6th Ed.). John Wiley & Sons. 2006. ISBN-13: 978-0471457282.

[8] Hedbrant, J. *On the Thermal Inertia and Time Constant of Single-Family Houses, Linkoping Studies in Science and Technology*, Thesis No. 887. Sweden: Institute of Technology, Linkopings Universitet, 2001. ISBN 91-7373 -045-9.

[9] James G. *Quintiere, Principles of Fire Behavior*. CRC Press, Taylor & Francis Group, 2017.

Ertuğrul, Nesimi, Personal photograph, "Dome-Shaped Oven Construction, Adelaide, Australia," November 2006.

Oven Design and Construction

3

3.1 INTRODUCTION

Wood-fired ovens have a rich history spanning centuries, with dome-shaped ovens emerging as a dominant and successful design. These ovens, characterized by their robust structure, have been tested and perfected over time. In this book, the intricacies of dome-shaped wood-fired ovens are explained, exploring their design and construction to create the optimal cooking environment.

The foundation of a dome-shaped oven lies in its two major sections: the hollow upper half of a sphere and a flat floor. This design offers an ideal shape from a structural integrity standpoint. The masonry sphere dome, with equal thickness throughout, provides perfect compression and ensures uniform heat distribution for cooking various foods.

While alternative designs, such as rectangular ovens with vertical brick walls, have been proposed, rectangular interiors within ovens hinder the desired direction of reflected radiant heat, especially during open-door cooking—a common practice in wood-fired oven cooking. The square or rectangular structure inhibits the efficient circulation of heat, leading to suboptimal results. Additionally, non-spherical structures with corners can create cool spots due to uneven air circulation, compromising the reliability of the oven. The need for additional structural support further diminishes the appeal of these alternative designs.

In this book, the true method of constructing dome-shaped wood-fired ovens is explained and the importance of the dome's shape in ensuring optimal heat distribution and cooking performance is explained. The historical success and knowledge accumulated over generations of dome-shaped oven builders are leveraged.

In the upcoming chapter, the practical aspects of wood-fired oven design and construction are explained. The heat distribution profile of dome-shaped ovens through the use of thermal images and measured temperature data will be given later to provide a comprehensive understanding of the benefits and intricacies of dome-shaped wood-fired ovens.

3.2 PLANNING, SPACE, LOCATION, AND SAFETY

Constructing a wood-fired oven from bricks requires careful planning and design considerations to ensure successful results and optimal cooking quality. The planning stage involves determining the appropriate location and designing the oven with careful attention to detail.

When planning, it is crucial to determine the maximum space required for the desired size of the wood-fired oven, whether it is for domestic or commercial use. For an ideal cooking platform, it is recommended that the inner diameter of a domestic dome should be at least 1.2 m. This size allows for a spacious fireplace and

DOI: 10.1201/9781032640136-3

ample room to cook multiple items simultaneously, such as flatbreads and cooking pots. In Türkiye it has been observed that some commercial wood-fired ovens have inner diameters of 3 m, which are continuously heated. Due to the long heating process required for wood-fired ovens, a relatively larger space is necessary to ensure effective and uniform heat transfer around the cooking area, optimizing the efficiency of heat utilization.

Two possible wood-fired oven layouts are illustrated in Figure 3.1, providing a top view that helps define the total space required for construction. As illustrated, in addition to the elevated slab to support the dome structure, a working surface adjacent to the door is essential (as shown in Figure 3.1a). Additional space on either side of the elevated slab may be considered for parking pre- and post-cooking items, as well as integrating a sink for dishwashing purposes (as depicted in Figure 3.1b), particularly for larger cooking utensils and tools.

A sea room in front of the oven is necessary for manoeuvring during cooking, although it may be limited for safety reasons, especially during party cooking. This can be addressed by using movable tables or benches, or by implementing a permanent restricted structure. These additional surfaces can be utilized for various baking activities, including dough preparation, ingredient storage, and cooling space for the finished baked items.

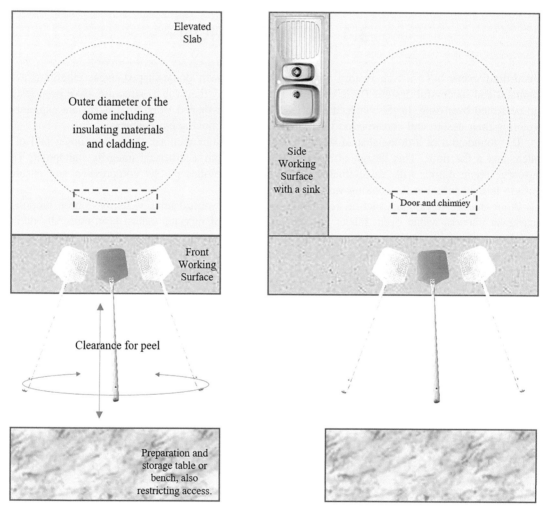

FIGURE 3.1   Top views of two possible oven layouts to define the space required (excluding the roof space). a) Top-down layout of the oven with an elevated slab designating the outer diameter. Below is the front working surface indicating the necessary clearance for a peel. This setup is complemented by a preparation and storage table or bench, which also serves as a restrictive barrier for access. b) Overhead view of the oven illustrating the side working surface equipped with a sink. The dashed lines outline the door and chimney's location for ventilation and access.

Ertuğrul, Nesimi, Personal drawing.

Proper planning and consideration of these spatial requirements and safety measures are crucial when constructing a wood-fired oven, ensuring a successful and enjoyable cooking experience.

It is important to note that the designs shown in Figure 3.1 are specifically for free-standing and outdoor wood-fired ovens. If these ovens are located within or attached to buildings, it is essential to carefully consider various safety measures. These measures include fire-proofing, fire prevention, and the prevention of possible smoke that may occur during the initial firing or when using inappropriate or high humidity firewood. While these safety precautions are also relevant for outdoor ovens, they are not as strictly required.

For optimal protection against environmental effects, it is recommended to construct an overarching roof over the dome of the wood-fired oven. Additionally, if space allows, a larger roof adjacent to a backyard pergola is highly recommended. This larger roof provides a comfortable working area that offers shelter from rain and sun.

In general, when defining the space and location for an outdoor wood-fired oven, several key considerations and safety precautions should be considered. These include:

- Allowing for a large enough space, referred to as the "sea room," to comfortably position the oven tools during cooking.
- Ensuring there is a sufficient surface in front of the oven door, not wider than 0.5 m to optimize the length of the peel handle. This surface can be used to place pans, pots, and other cooking utensils during the cooking process. Additionally, incorporating a side storage and working surface with a sink for dishwashing purposes is advisable.
- Positioning a sufficiently large table or bench in front of the oven, which can also function as a barrier. This surface can be used to hold a dough ball tray (as shown in the next chapter), cutting boards, rolling pins, and filling ingredients. It can also serve as a preparation area and provide space for cooling down baked or cooked food. It is worth noting that in commercial ovens, this space can be as large as the oven itself.
- Ideally, for domestic applications, situate the wood-fired oven adjacent to an entertaining area that allows guests to witness the cooking process. This provides them with an immersive experience, engaging their senses of smell and sound.
- For outdoor versions, it is important to have easy access to a nearby hose connection for space cleaning and regularly replacing water in the cleaning mop bucket.
- When designing the wood-fired oven, consider the orientation of the dome door in relation to the nearby house and prevailing wind. It may be necessary to use a partially closed door design to mitigate the effects of the prevailing wind.
- Even for outdoor ovens, it is essential to follow fireplace construction regulations for safety purposes.
- A granite-top front space is recommended for its durability and ease of water-based cleaning. However, it is important to note that if the granite top is too thin (around 2–3 cm), it may crack when hot clay pots or stainless-steel cooking utensils are placed on it for an extended period.

By considering these factors and implementing the necessary safety precautions, an enjoyable and safe cooking environment for an outdoor wood-fired oven can be created.

3.3 FEATURES AND DESIGN ASPECTS

While various dome structures have been used for wood-fired ovens, two main designs are distinguished based on the location of the firing place: directly fired ovens (where the fireplace is on the same level as the food to be cooked) and indirectly fired ovens (where the fireplace is at the basement level, allowing for slow cooking and better utilization of the cooking surface).

Figure 3.2 illustrates these two distinct wood-fired oven structures, which are applicable to both indoor and outdoor types. These structures primarily differ based on the integration of the fire source and

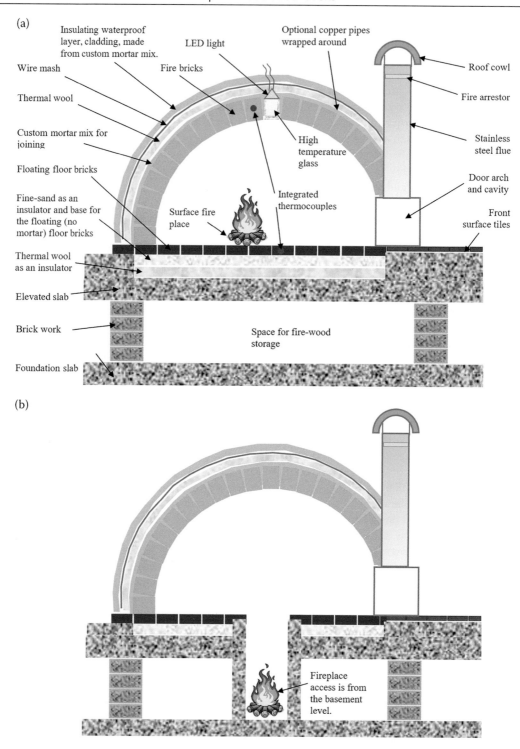

FIGURE 3.2   Construction details of two major wood-fired oven structures: a) The most common oven type with a surface-level fireplace, internally fired and b) An oven with a fireplace in the basement, which can be referred to as an externally fired, indirectly fired, or hidden-fired oven.

Ertuğrul, Nesimi, Personal drawing.

the flue path. The cross-sectional drawings in Figure 3.2 demonstrate the major materials used and their positions in the main structure of the oven. These traditional wood-fired oven structures are also referred to as internal combustion ovens, retained-heat ovens, or direct-fired ovens.

Although it may not be a commonly used design, the oven structure given in Figure 3.2b is known as the "white oven" and is frequently utilized in commercial bread ovens in the western and northern regions of Türkiye. As mentioned earlier, this design offers a larger and cleaner cooking surface, making it suitable for various Turkish bread types such as Trabzon bread, round/long loaves, somun breads, prina bread, as well as pumpernickel and sourdough breads. Figure 3.2b showcases a fire chamber located below the cooking chamber, where the heat from the fire is vented into the cooking chamber through a small hole in the middle of the oven. This design maximizes heat utilization. However, it should be noted that this externally fired and vented oven structure is more complex and expensive to construct compared to the internally fired design.

Both oven structures bake through a combination of retained heat and continuous heat, making them highly efficient for baking large quantities of bread compared to a tandoor oven structure. However, the design shown in Figure 3.2b is not only more efficient than Figure 3.2a, but also offers a larger net floor space. It is worth mentioning that if the fire is maintained on one side of the oven chamber in the design shown in Figure 3.2a and if it is well insulated, this retained-heat oven requires less masonry mass.

In certain oven building practices, it has been found that the floor of the oven can be inclined a few degrees towards the door. This inclination can make it easier to load and unload the oven with a peel, and it can also help retain steam inside the dome during baking.

In both designs, heat energy generated by the fired wood is transferred through conduction in the walls of the dome and floor, and through convection as cold air enters the oven through the top section of the door cavity. Convection currents then transfer the heat energy inside the dome cavity and exit from the upper cavity of the door, as illustrated in the previous chapter in Figure 2.3. Both designs require a sufficiently high thermal mass to achieve a high heat capacity.

In the oven type shown in Figure 3.2b, the absence of firewood and ashes on the oven floor next to the cooked item simplifies maintenance and keeps the floor cleaner for longer periods. In the western part of Türkiye, specifically the Aegean Region, pirina is commonly used as a fuel source for such ovens. Pirina consists of dried olive seeds, which are a recycling product from olive processing plants. In other regions, alternative low-grade new or recycled wood and plant products are used as fuel in these ovens.

An additional highly beneficial design feature can be incorporated to enhance the effective utilization of generated heat. This is illustrated in Figure 3.2a, where thin copper pipes surround the dome structure and are positioned under the insulating materials. These continuous copper pipes can circulate water to capture heat energy for water heating or for space heating through a radiator located inside a nearby residential or commercial space.

The characteristic features and design aspects of wood-fired ovens will be described in six main sections below, which are closely associated with and follow the same order as the construction steps. These construction steps will also be followed in the explanation and demonstration of two sample ovens, which will be provided at the end of this chapter under the subtitle "Building Stages of Two Sample Ovens."

3.3.1 Foundation (Ground) Slab, Walls, Elevated Base Slab and Tools, and Wood Storage

Figure 3.3 provides an outline of the oven components that need to be considered during the building process. The space under the elevated base slab can be utilized for storing cooking tools and firewood.

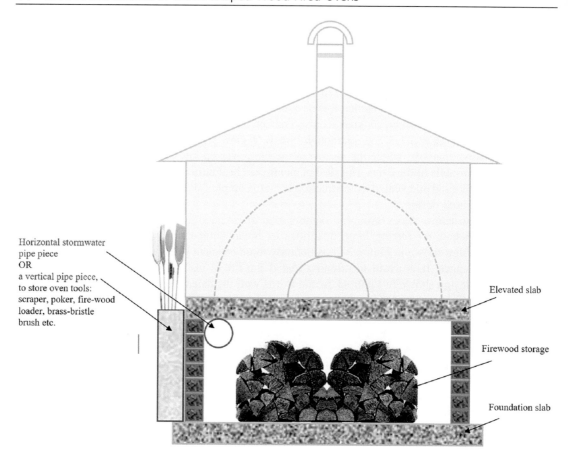

FIGURE 3.3   Outline drawings highlighting the components to be considered in the first phase of the wood-fired oven building.

Ertuğrul, Nesimi, Personal drawing.

Some wood-fired oven builders have suggested integrating a rectangular hole in the front of the oven door and on the front section of the base slab, which can be used to remove ash through a built-in drawer. However, this book does not recommend this approach as it adds complexity to the structure without offering significant benefits. There is no need to remove the ashes after every cooking session. Instead, the ashes can be collected and used as a barrier to reduce flame radiation during cooking.

As previously mentioned, outdoor wood-fired ovens may require a separate roof to protect against rain and sunlight. While the cladding on "the dome-only structure" may provide a weatherproof exterior, direct exposure to temperature extremes and weather conditions can lead to degradation and the development of cracks. Therefore, it is advisable to include a roof in outdoor design.

3.3.2 Features of the Floating Brick Structure

The design of the oven floor requires special attention to address three major functions:

- Providing a sturdy base to support the dome structure.
- Offering a wear-resistant and food-safe surface as various food products come into contact with the floor bricks during the cooking process.
- Allowing for efficient conduction heat transfer.

As illustrated in Figure 3.2a, the floor bricks within the fire chamber, as the main cooking surface, should minimize heat loss and possess thermal inertia to sustain conduction heat over extended periods. This can be achieved by using appropriate brick types and a sufficiently thick layer of fine sand (approximately 50 mm) as an under layer. The fine sand also serves the purpose of securing the floor bricks in position and expanding and contracting together during heating, while facilitating easy replacement of damaged floor bricks.

The lower layer, situated above the elevated slab and below the floor bricks, should provide effective insulation. Various materials can be used for this under layer, depending on cost, availability, and effectiveness. Common practical materials include rock salt, finely crushed recycled glass pieces, raw iron/ore powders (often used in the Divrigi region of Türkiye, known for its rich iron ore mines), and high-temperature thermal wool, which is recommended in this book due to its desirable thermal characteristics.

While firebricks are robust when in direct contact with fire, their porous structure makes them unsuitable for the floor where direct food contact occurs. Additionally, soapstone, which has excellent thermal properties and is commonly used for making cooking pots, bowls, and slabs, should be avoided on the floor structure due to its primarily talc composition, which makes it soft. During baking and cooking, liquids and ingredients often drop onto the floor surface, requiring regular cleaning, typically with a metal brush.

An important characteristic to consider for the stone used on the floor is its heat absorption rate and density. A denser rock requires less material to hold the same amount of heat energy. Therefore, stones with higher energy density (specific heat multiplied by density) have a greater ability to absorb heat for a given thickness or size, making them suitable for the floor of a wood-fired oven.

In conclusion, the hardness of the floor bricks is a key requirement for healthy and reliable cooking. Depending on availability, a durable floor can be constructed using various natural stones, such as basalt. Hard-wearing, high-temperature paving bricks are recommended for forming a low-cost floating floor structure as they can withstand the weight of the dome and temperature variations. These temperature variations, which can reach up to 20°C, occur regularly when mopping the floor with cold water for cleaning and temperature regulation.

It is important to note that the number of floor bricks or pavers required for a given oven can be calculated by knowing their dimensions and the size of the slab.

3.3.3 Dome Features and Extras

The dome structure is a critical component of a wood-fired oven as it controls and contributes to all methods of heat transfer and provides a large thermal mass necessary for various cooking techniques. While hard red bricks can be used, they are not the optimal choice for high temperatures, flame exposure, and frequent thermal cycling. Instead, firebricks are recommended for the dome structure. To reduce costs and weight, it is advised to cut standard size firebricks (230 × 115 × 63 mm) in half, resulting in bricks with dimensions of 115 × 115 × 63 mm. The effectiveness and validity of using such cut firebricks in the dome structure have been tested and accurately measured in this book, as detailed in Chapter 5. The results indicate that with suitable insulation processes, as illustrated in Figure 3.2, a sufficient thermal mass can be achieved for both domestic and commercial wood-fired ovens.

The dome bricks are set and mortared together in a dome shape, with a line of mortar between and behind them, so that one of the uncut edges of the brick faces inside the oven chamber. This arrangement results in a brick-only dome thickness of approximately 115 mm, with an additional thickness of about 40–50mm due to mortar, wire mesh, and thermal insulation.

Before constructing the dome structure, it is necessary to calculate the number of firebricks required. This can be easily done by determining the area facing the oven chamber, which is half of the sphere's surface. For example, using the given standard bricks with a brick face of 73 cm² and a dome with an inner radius of 60 cm, the number of bricks required would be $(4\pi r^2/2)$ /73 cm² = 2 × 3 × 3.14 × $(60)^2$/73

= 310 half bricks and 165 full bricks, minus the area needed for the door arch. If uncut bricks are used and mortared into a similar arched vault, 310 full bricks would be needed. It is recommended to have approximately 10% more bricks for both the dome and the floor work to account for any building faults or handling mistakes.

The final thickness of the dome can be determined by adding thermal wool insulation, wire mesh, and an additional cladding of about 30 mm thickness, as illustrated in Figure 3.2a. In addition to the outer insulating waterproof layer of 30 mm, further cladding may be required for a better finish, such as covering the dome with decorative miniature tiles, if there is no outer wall in the design and only the dome is visible.

During the building phase, thermocouples can be integrated by inserting them inside the half-drilled dome and floor bricks, as shown in Figure 3.2a. Sufficiently long thermocouple wires should lead to signal conditioning and temperature indicator circuits or be connected to a data logging system, as practiced and presented in Chapter 5. Thermocouple-based temperature measurement, although slightly costlier than handheld devices, is recommended as it provides continuous observation of the oven temperatures at the correct locations.

In addition, while it can be enjoyable to watch the wood fire flames during the firing process and while roasting certain food items, the absence of flames (to eliminate flame radiation) is preferred in the majority of cooking practices. However, this absence of flame can make visibility inside the oven poor or impossible, particularly at night. To address this issue, it is recommended to integrate a light source during the construction of the dome. This can be achieved by embedding a high-temperature glass, such as a heatproof square-shaped Pyrex bottle with a thickness similar to that of the dome wall, into the dome structure, as shown in Figure 3.2a.

The construction of the dome is often seen as highly challenging brickwork by many. However, this process can be greatly simplified by using a hybrid formwork made of plywood and thick cardboard strips. While different dome-building techniques are employed in practice, this method is suggested here due to its cost-effectiveness and its ability to facilitate the solidification of brick and mortar combinations into the desired sphere shape. Using this method, the construction time of the dome can be reduced to a single day. The principal plywood formwork can be created by cutting out multiple quarter-circles (approximately 12 or more) and joining them together at the middle, as shown in Figure 3.4a. These shapes can then be covered with cardboard strips to form a structurally sound half-sphere shape, as shown in Figure 3.4b. This formwork allows for easy alignment of the dome bricks, ensuring that their connecting surfaces touch as closely as possible, thereby avoiding any mortar dropping inside the chamber over time (see Figure 3.4c).

(a)   (b)   (c)

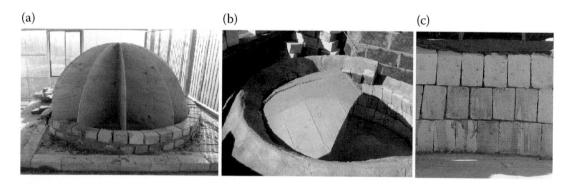

FIGURE 3.4   Building a dome using a plywood formwork (a, b), and joining dome bricks, no mortar between the bricks (c).

Ertuğrul, Nesimi, Personal photograph, "Dome-Shaped Oven Construction, Adelaide, Australia," November 2006.

Ideally, the inner faces of the firebricks, which face the cooking and firing surfaces, should touch each other, while the back section of the dome should be joined using the correct mortar mixture. It is important to note that gaps between the inner faces can lead to loose mortar dropping onto the cooking item, and the use of the wrong brick type, such as red wall brick, can result in gradual decay of the brick material. Additionally, it is not recommended to insulate the dome with loose fibre insulation such as rock wool or fiberglass. As illustrated in Figure 3.2a, the first layer of mortar on the back of the fire-bricks, along with thermal wool and a holding mortar, has been found to be a highly effective insulator, as also verified during the data logging processes.

Whether an indoor or outdoor wood-fired oven, the structural design remains mostly consistent, except for the addition of an outer wall in some cases, which may be constructed at the edge of the elevated slab. As previously mentioned, adding a roof can help protect the dome structure from external influences such as rain and sunlight.

It should be noted that building the entire oven structure using clay has been practiced but is not recommended due to several drawbacks. Clay is water-soluble, lacks good insulation properties, and can wear away easily. Inadequate cladding and the use of the wrong mortar mixture can also lead to the formation of cracks in the dome structure, particularly under thermal cycling. Achieving the desired temperature in the oven structure is crucial, as inadequate heat transfer or a small thermal mass may require over-firing, which can result in cracks.

Proper selection of materials for the dome and floor, along with effective insulation, ensures relatively equal heat transfer from the dome walls and the floor. This is essential for achieving uniform and consistent cooking results, especially for dough-based products. For example, when cooking pide bread or lahmacun, the ideal cooking temperature for the floor is between 270–290°C (primarily achieved through conduction heat transfer), while the temperature of the dome wall is between 300–350°C (primarily achieved through flame-based heat transfer, with the reflected heat reaching the cooked items).

It is recommended to avoid using baking stones (which are poor heat conductors) or thin metal trays when cooking dough products, as they can disrupt the heat conduction characteristics of the floor, leading to undercooked inner layers or overcooked top layers. Additionally, when baking certain bread types to achieve a better crust, it is necessary to introduce additional steam into the oven. This can be done by adding water to a tray or using firewood with higher humidity content. The quality of the bricks used is also important, as humidity can affect their integrity and, consequently, the lifespan of the oven.

3.3.4 Door Frame, Optimum Door Size, and the Door Itself

The floor, dome, and door are crucial components of a wood-fired oven that are prone to heat loss. Proper understanding of heat losses associated with these sections is necessary to ensure optimal temperature control and cooking results.

The door of the oven plays a significant role in regulating heat and minimizing heat loss. The selection of an appropriate door type (as shown in Figures 3.5a and 3.5d) and its size for open-door and partially open-door cooking is important. The door panel should be designed to provide effective thermal insulation, particularly for slow cooking that requires extended cooking durations.

When determining the dimensions of the oven door, it is recommended to maintain a ratio between the height of the door and the height of the oven dome of approximately 2/3 to 1/2 (=1.33). This helps to avoid excessive heat loss, particularly during open-door cooking.

Wood-fired oven doors come in various shapes and sizes, as shown in Figure 3.5. The size (area) of the door should meet two main criteria:

- It should be wide enough to allow the largest cooking utensils to pass through easily. Typically, the width of the door is determined by the widest stainless-steel tray commonly used for slow cooking and roasting, while the height is determined by the tallest cooking utensil, such as a tajine with a tall lid.

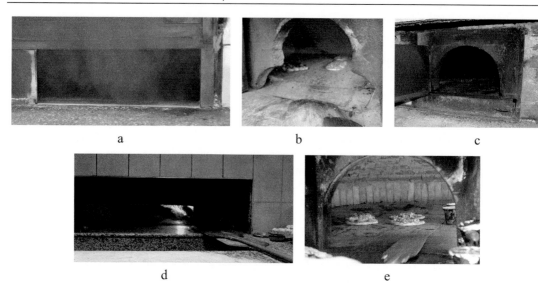

FIGURE 3.5 The wood-fired ovens commonly feature various types of door structures to meet different cooking needs: a) Rectangular door with a sliding metal door panel, which allows for easy opening and closing of the door, providing convenient access to the oven chamber, and helps regulate heat and minimize heat loss during cooking; b) Nested/varying size half-circle shapes without a door panel that is commonly used in continuous bread cooking in commercial ovens, and the half-circle shapes create separate air inlet and outlet areas, promoting proper air circulation within the oven; c) Half-circle shape with a hinged metal door panel, which features a door panel that is hinged to the oven structure, allowing for easy opening and closing, while providing a reliable seal to prevent heat loss and control the cooking environment; d) Long/thin rectangular shape door with a sliding door panel that is typically used in larger wood-fired ovens, particularly in the oven types shown in Figure 3.2b, which also allows for convenient access to the oven chamber and efficient heat regulation; e) Rectangular/half-circle combination door with hinged or free-standing door panel that incorporates both rectangular and half-circle elements, offering a versatile and customizable door structure that features a hinged door panel or a free-standing door panel, providing flexibility in terms of access and heat control.

Ertuğrul, Nesimi, Personal photographs, "Dome-Shaped Oven Doors, Türkiye," April-May 2010.

- The door area should be sufficient to allow proper air movement, which is crucial for preventing smoke from escaping through unintended openings, particularly in indoor ovens.

In Figures 3.5a and 3.5d, the rectangular door frames feature partially closed openings achieved through sliding door arrangements. Partially closed doors are commonly used in commercial ovens once the desired heating level is achieved. This helps reduce heat dissipation and influences convection heat transfer by limiting external air movement. The lower open section of the door allows fresh cold air to sweep the floor surfaces, which can have a significant cooling effect, especially in cold ambient temperatures. Additionally, as mentioned in Chapter 2 and Chapter 5, proper sealing and closure of the oven door are essential for slow cooking.

By considering these factors and optimizing the design of the door, heat losses can be minimized, leading to more efficient and controlled cooking in the wood-fired oven.

The design of the door opening in Figure 3.5b incorporates a unique feature that helps capture and discharge smoke effectively. The smaller top half-circle section serves to collect and channel the smoke towards the chimney, while the larger bottom section allows for easy access to the oven chamber and provides better visibility inside. It is important to ensure that the door opening is adjusted correctly to maintain optimal heat control during cooking. In cases where heat control is not sufficient, techniques such as wetting the floor with a mop or removing ignited wood can be employed.

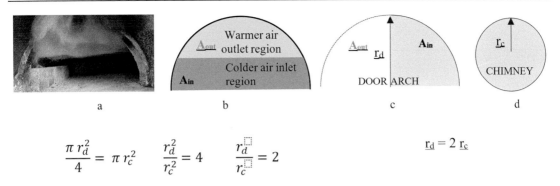

$$\frac{\pi \, r_d^2}{4} = \pi \, r_c^2 \qquad \frac{r_d^2}{r_c^2} = 4 \qquad \frac{r_d^{\square}}{r_c^{\square}} = 2 \qquad\qquad r_d = 2 \, r_c$$

FIGURE 3.6 The relationship between the area of the door cavity and the chimney diameter ($A_{out} = A_{in}$), where r_d is the radius of the door arch and r_c is the radius of the chimney.

Ertuğrul, Nesimi, Personal drawing.

In a wood-fired oven, the movement of air is dictated by the laws of physics. Heated air naturally rises within the dome, carrying smoke (unburned wood particles) along with it, while cooler and denser air moves into the oven. It is crucial to direct the smoke fully towards the chimney inlet to prevent the accumulation of dark residue outside the door arch, which can lead to undesirable air pollution in the vicinity. The size of the door and the corresponding chimney can be determined by considering the relationship between the area of the cold-air inlet and the area of the warm-air outlet.

The optimal chimney size for a given door can be determined using the formula shown in Figure 3.6. The formula indicates that the area of the cold-air inlet should be equal to the area of the warm air/smoke outlet (as given in Figures 3.6a and 3.6b). By ensuring that the chimney radius is greater than half of the radius of the door arch ($r_c > r_d/2$), the smoke will flow through the chimney cavity, preventing the build-up of dark residue outside the door arch. This design approach can be applied to different types of door structures as well.

It is important to note that smoke during firing may occur before ignition or when using improperly dried wood or wood with chemical substances, such as paint. Taking necessary precautions in wood selection and ensuring proper drying can help minimize the occurrence of smoke and optimize the performance of the wood-fired oven.

There are various door-panel structures commonly used in wood-fired ovens, including sliding metal types with single or two sections, conventional metal door panels on hinges, and free-standing metal types. Figure 3.7 illustrates the cross sections of two different door and door panel designs, as well as their integration with the chimney and flue tube sections.

A well-designed door-panel combination should have a proper seal and thermal insulation, along with an insulated handle. In the design shown in Figure 3.7a, the door frame consists of two different arch structures, accommodating the door with multiple layers of insulation (including a metal front panel, an air pocket, and a wooden back panel). When the door is closed, it provides a tight seal for the entire dome chamber using a high-temperature fireproof sealing rope. This design is recommended for wood-fired ovens as it ensures a proper seal and minimizes heat loss during slow cooking.

The design in Figure 3.7b offers a simpler and more cost-effective structure, which is commonly used. However, the free-standing metal door plate in this design has poor insulation characteristics, making it less suitable for long-hour slow cooking practices. In some door-cavity designs, the flue inlet is positioned inside or outside the doorway. The inside version usually incorporates a flue tube shutter, as illustrated, which aims to provide sealing and insulation when used together with the free-standing L-shaped metal door plate. However, this design is not recommended due to significant heat loss.

Therefore, it is recommended to position the chimney in front of and above the oven door (as shown in Figure 3.7a), preserving the sealed integrity of the cooking chamber inside the dome and retaining heat when the active fire is removed for slow cooking. Additionally, as mentioned earlier, the door panel should be thermally insulated and airtight to prevent steam from escaping during slow cooking and certain baking practices that require moist and crisp bread.

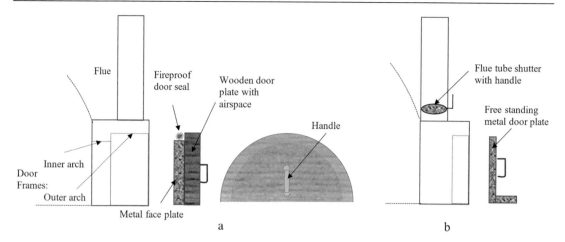

FIGURE 3.7   The cross sections of two door design options and door panels (not scaled): a) "A flue above/outside the oven doorway design" with a sealed door (recommended). b) "A flue above but inside the oven doorway design" with a flue tube shutter and a free-standing metal door plate.

Ertuğrul, Nesimi, Personal drawing.

In some door designs, hinged metal doors are used to limit air movement during burning. A half-closed door panel can also help prevent draft air during open-door cooking. Therefore, hinged doors are often made with two sections to allow for control of draft air (as shown in Figure 3.5d). In certain commercial ovens, counterweighted doors are employed, which can either fall away in front of the peel or be manually lifted up and out of the way.

Furthermore, due to its simplicity, both designs in Figure 3.7 utilize an L-shaped metal plate with a wooden backing (or metal plate only) and a D-handle on the outer side. However, the insulation characteristics of the Figure 3.7b design were found to be ineffective and inefficient in slow cooking, as the door panel sits inside a single arch structure and results in significant heat loss.

3.3.5 Chimney, Flue Tube, and Fire/Spark Arrester

In a wood-fired oven, when firewood is burned in the oven chamber, the exhaust exits through the oven door and up into the flue and chimney located just above and outside the oven door. The chimney section of a wood-fired oven has a crucial safety duty: effectively channelling the warm air and smoke from the fire chamber to the outside of the oven, providing sufficient suction, and preventing potential fires in nearby areas.

There are two practices that should be avoided when it comes to chimney and flue tube placement:

- Placing the chimney/flue tube at the back, in the middle of the dome, or inside the oven doorway, as this can result in excessive heat loss and inadequate heat retention even during continuous wood burning.
- Firing wood at the door cavity or in the middle of the cooking area, which also leads to significant heat loss and unnecessary cleaning work.

A chimney cowl (cap) should be attached to the top of the sufficiently long flue tube (as shown in Figure 3.8a). This is necessary not only for fire safety, as a taller chimney provides stronger draft (as illustrated in Figure 3.2a), but also to keep out rainwater, hail or snow. A long, straight flue tube also helps mitigate the impact of wind and improve suction. It is worth noting that the design presented here may not prevent bird nests, but additional metal spikes on top of the chimney cowl can be employed to address this issue.

a   b

FIGURE 3.8   Images of two critical components of a flue tube: a) Stainless-steel cowl, anti-draught chimney rain hat; b) Spark arrestor that fits inside flue tube to restrict the emission of sparks and embers.

Ertuğrul, Nesimi, Personal drawing.

In large urban bakeries or busy restaurants, fires are active during long baking or cooking sessions. Similarly, although less frequent, long-hour cooking may also occur in domestic wood-fired ovens. Therefore, in addition to a properly functioning chimney and flue system, a spark arrestor is essential to prevent flammable debris from escaping the combustion chamber and landing on the roof or combustible materials, which could potentially cause a fire.

The spark arrestor consists of double layers of metal mesh (as shown in Figure 3.8b), which can capture embers while allowing gases and smoke to exit the chimney. It is important to remember that spark arrestors require regular cleaning based on the frequency of firing and the types and quality of firewood used.

3.3.6 Flame Radiation Barrier and Ash Reservoir

As discussed in detail in the previous chapter, flame radiation is not the preferred heat transfer mechanism in many wood-fired oven cooking techniques. Some wood-fired ovens employ methods such as starting the fire in the middle of the floor and shifting it to one side before cooking, or using a metal wood stand to keep the active fire away from the items being cooked. However, both of these methods are not recommended in wood-fired oven practices.

Starting the fire in the middle of the floor and shifting it can accelerate burning and produce intense flame radiation, which causes heat loss as well as leads to overcooking of the top surface of the cooking substance. Similarly, using a metal wood stand can also result in excessive flame radiation and adversely affect the cooking process.

Instead, it is recommended to allocate a dedicated fireplace inside the dome using a barrier made of a line of red bricks or a U or L-shaped thick metal bar (as shown in Figure 3.9 and Figure 2.3 in Chapter 2).

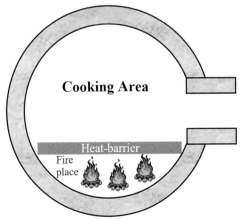

FIGURE 3.9   The cooking area and the fireplace divided by the heat barrier (bricks, metal bars, or ash-hill).

Ertuğrul, Nesimi, Personal drawing.

This barrier, commonly used in commercial wood-fired ovens in Turkiye, is typically around 10–14 cm in height. In the large commercial ovens, the barrier can be formed by the accumulated ashes (ash-hill) during continuous wood fire burning.

The heat barrier serves two main purposes: minimizing flame radiation and providing an ash reservoir to avoid frequent ash removal and keeping the cooking surface free from fire debris during both initial and continuous firing. The active fire should always be kept behind the heat barrier to maintain proper heat distribution within the oven.

3.4 BUILDING MATERIALS

As mentioned earlier, a wood-fired oven consists of a floor and a dome, both constructed using bricks, and both structures need to be properly insulated for efficient heat utilization. When thermal insulation is done correctly, heat loss is minimized, and the oven maintains a stable temperature range necessary for effective cooking. Other important components include a well-designed door structure connected to a chimney and a baking chamber, with an acceptable ratio of door size to the overall oven.

There are several features and characteristics to consider when building a wood-fired oven. The size of the oven, primarily determined by the internal diameter of the dome, is an important consideration. It is worth noting that the geometry of a dome-shaped oven means that the radiant heat reaching the cooked items is relatively independent of the oven size. Additionally, due to the space required for the fire, the usable cooking surface area will be less than the total circular floor area. For domestic use, a smaller or medium-sized oven is often sufficient, as a full load during cooking helps maintain moisture levels, and a smaller oven can bake just as well as a larger one.

If the primary goal is to cook pizzas, a wider and thinner dome design with lower thermal inertia is acceptable, along with a wider door. These factors influence the amount of material required to build the oven. On the other hand, if loaf bread is the main focus, a similar dome structure with a smaller door is ideal. It is worth noting that it can be challenging to maintain an active fire while cooking a series of pizzas in a very small oven.

In addition to the bricks used for the floor and dome, other primary construction materials include concrete for the foundation and elevated slabs. Special mortar is required for the dome, and insulating materials are needed for the base and dome. Figure 3.2 illustrates the typical locations of these materials in an oven structure. Table 3.1 provides specific heat capacities of various materials, offering a comparison of their characteristic parameters. It is important to note that the specific materials' heat conductance and capacity may vary depending on the manufacturer, local conditions, and moisture content. Materials with low thermal conductivity are suitable for insulation, while those with high specific heat capacity are useful for thermal storage. Since both of these characteristics are desirable for the dome and floor constructions, there are conflicting demands when selecting construction materials.

For the construction of a wood-fired oven, multiple materials are used in the dome structure, such as bricks, cement-lime mortar for structural integrity and heat retention, glass wool for insulation, and wire mesh with cement-lime mortar on the outside. The integration of these materials into the final design has a significant impact on the heat-retaining capacity and thermal mass (thermal inertia) of the oven.

The choice of bricks used in the dome and floor is crucial for achieving the desired thermal mass. High-grade, high-density, and lightweight firebricks are recommended for the dome construction, as they can withstand the cycling heat and do not transfer excessive heat to the food during ideal cooking

TABLE 3.1   Characteristics parameters of some materials that can be used in the wood-fired oven construction

MATERIALS	SPECIFIC HEAT CAPACITY (KJ/KG.K)	THERMAL CONDUCTIVITY (W/M.K)	GROSS DENSITY (KG/M³)
Brick 1500 kg/m³	0.85	0.40	1500
Brick 700 kg/m³	0.85	0.17	700
Brickwork	0.82	0.73	2200
Brickwork	0.92	0.16	800
Mortar	0.80	0.93	950
Cement mortar	0.80	1.40	2000
Cement-lime mortar	1.10	0.80	1800
Lime-cement mortar (for masonry)	1.00	1.05	1800
Clay/Silt	1.50	1.50	2000
Sand/Gravel	1.00	2.00	2000
Concrete 2400 kg/m³	1.00	2.00	2400
Tiles, clay	0.80	1.00	2000
Tiles, concrete	1.00	1.50	2100
Glass (quartz glass)	1.05	1.38	2200
Glass wool	1.03	0.04	50
Iron, cast	0.45	50.00	7500
Stainless steel	0.46	17.00	7900
Steel	0.45	50.00	7800
Roof tile (clay)	0.80	1.00	2000
Roof tile (concrete)	1.00	1.50	2100

conditions. On the other hand, as stated previously, low-density bricks should be avoided for the floor structure, as they have direct contact with the food and dough products.

When selecting floor bricks, three criteria should be considered: food safety (direct contact with food), minimal heat transfer to the bottom of the food, and durability under cycling temperature variations (especially during the cleaning process under an active fire).

It is important to note that dome bricks primarily contribute to radiated heat transfer, while floor bricks primarily contribute to conductive heat transfer. However, during cooking, it is desirable for both the dome and the floor to transfer heat to the food in a balanced and moderate manner.

Standard and low-cost bricks should be avoided in the oven construction, as they are more likely to crack, spall, and deteriorate over time due to cycling heat. In some European countries, specialized materials are manufactured specifically for oven use, offering higher resistance to thermal shock than standard firebricks. High-temperature paving bricks are recommended, as they are vitrified and have better heat resistance, maintaining their shape and durability.

While high-temperature bricks may be considered brittle, using the floating assembly method (placing them over a layer of fine sand) helps prevent cracking. High-temperature paving bricks provide better resistance to thermal cycling while conducting heat at a slower rate. It is important to remember that the floor bricks are regularly cleaned with a wet mop and scrubbed with a metal brush, causing rapid temperature changes, which can lead to the breakdown of the brick material. If available, gently rounded-end bricks can be considered for the floor structure, as they are more resistant to chipping.

TABLE 3.2 Ratios in volume for each mortar that can be used in the oven dome construction

RATIOS IN VOLUME		
LIME PUTTY	CRUSHED ROCK POWDER	SAND
1	–	3
1	1	2

3.4.1 Mortar Mixture

As mentioned earlier, it is important to avoid using metal reinforcement in the wood-fired oven structure due to its different rate of expansion and contraction compared to masonry. Pure masonry cement also has limitations, including slow strength acquisition during setting and a decrease in strength with higher water content or premature drying. Although plain masonry cement can tolerate an average temperature of 300°C, it becomes weaker under long-term heat and repeated cycles, which are typical in wood-fired ovens. To overcome these limitations, it is recommended to use a suitable mortar mixture for the dome construction.

A common mortar mixture for binding dome bricks while increasing adhesion and flexibility is a combination of sand, cement, and lime. Adding powdered clay to the mortar mixture can increase heat resistance but significantly reduces its strength. Studies have shown that increasing the lime content by 10% leads to a 14% decrease in mechanical strength of the binder and a corresponding 12% decrease in stiffness at 7 days of curing age [1,2]. Therefore, the ratios provided in Table 3.2 are acceptable for creating a mixed mortar with optimal properties.

Lime is one of the oldest binders to be used in construction and was used by ancient Greeks to produce masonry by mixing with sand, and mortars with lime putty showed better material properties [3]. Different types of limes were studied in [4]. After 28 and 90 hardening days, higher compressive and flexural strengths were achieved in lime. Lime putty coatings also demonstrated resistance to shrinkage cracks on the surface.

Note that calcium aluminate cement (which is expensive and may be difficult to find) can be used instead of common masonry cement, but it should be careful when handling it as flash-setting can occur during mixing.

It is recommended to use high-temperature (below 1200°C) insulation blankets (thermal wools) both in the floor and dome construction of the oven. The same material can be used in the chimney/flue tube insulation in indoor oven structures. In insulation blankets, it is desirable to have the following characteristics: thermal stability, retention of their structure without emitting fume or smell and low heat storage. In addition, it is desirable to be able to cut and handle the blanket material easily. Furthermore, insulation blankets should not to cause any skin irritations (as in traditional ceramic or fibreglass blankets) and should not present health risks. Note that there are alternative low-cost insulation materials with low thermal conductivity. Although the air pockets inside some of these materials can provide a good insulation their availability and handling may present difficulties.

The desirable characteristics of the bricks used on the floor and in the dome structure are described previously, hence will not be repeated here. As a reference, the list of material is given in Table 3.3 which can be used to build a domestic level wood-fired oven with an inner dome diameter of 1.5m. The building stages of this size oven have been provided the next section. It should be emphasized here that although the list in the table can offer a rough calculation to determine the material required to build a specific size oven, alternative oven designs (with aesthetically and architecturally different designs) will require additional materials.

TABLE 3.3 The material list for an outdoor domestic wood-fired oven with an inner radius of 0.75 m

MATERIAL	REMARKS
1 unit cement 2 unit lime 4 unit sand	The mortar mixture is necessary for constructing the dome. The total volume of mortar needed must exceed 5 cm (for building) + 3 cm (for outer cladding) x 2 × 3 × 3.14 x (75²) = 283,000 cm³.
Fine sand	This can be obtained from garden suppliers. The volume of the sand required: 5 cm (thickness) x 180 cm x 180 cm = 162,000 cm³
Wire mash	Wire mesh is required to cover the entire dome (half of a sphere), which amounts to approximately 50,000 cm² (including the door area). The mesh should have holes that are around 2cm x 2cm in size, which is sufficient to secure the insulating fire blanket and outer cladding in place.
High-temperature fire blanket for insulation	This material is required for insulating the floor and the top of the dome. In the sample designs illustrated in the following sections, Superwool 607, with a density of 96 kg/m³, has been used as an insulating fire blanket. The dimensions of this material in a standard size package are 2.5 cm in thickness, 732 cm in length, and 61 cm in width. The total area of fire blanket required for the dome and the floor is (180 cm x 180 cm) + (50,000 cm² for the wire mesh) = 83,000 cm² (equivalent to two standard size packages).
Flue kit: Flue tube, cowl and fire arrestor	The width (diameter) of the flue tube should be at least 20 cm or larger. The length of the flue tube should be sufficient to ensure proper safety and clearance, considering the total height from the bottom to the roof top, and a cowl and a fire arrestor should be appropriately fitted to the flue.
High-temperature paving brick for the floor.	The paving brick should have a smooth, heat-resistant surface. The red bricks used for the floor should cover an area determined by the outer diameter of the oven. For example, a square area of approximately 200x × 200 cm is needed, totalling 4 m². To calculate the number of bricks required, divide the total surface area (40000cm²) by the area of a single brick (242 cm²), resulting in approximately 165 bricks.
Firebricks	As recommended earlier, the standard size firebricks (230 x 115 x 63 mm) can be cut in half and used in the dome construction. It is preferable to use firebricks in a long trapeze shape as they require less mortar and provide better structural stability.
A brick cutter	Cutting disk with water should be preferred to avoid dust.
The light system for the dome	This option requires a suitable high-temperature glass to be integrated into the dome structure.
Heat barrier	A line of thicker red bricks (such as 220 x 110 x 80 mm) or an L-or U profile thick metal piece can be used. Its length should be about 1m for the floor size of 1.5m diameter.
Door panel	The door shape has few options as discussed previously. The door panel should suit to the door shape. However, it is recommended to have a properly insulated door plane that is vital for efficient slow cooking.
Cement required for the foundation slab	Standard concrete mix can be used, reinforced by thick metal mesh. Similarly, for the oven size considered here, total volume of the slab will be about 250 cm x 200 cm x 15 cm = 750,000 cm³.
Brickwork for the elevated slab	Standard ordinary wall bricks can be used for the construction. To determine the number of bricks required, the dimensions of the walls need to be known, including the desired height and the dimensions of the foundation slab.
Cement required for elevated slab	Standard concrete mix with reinforced by thick metal mesh can be used. Its size can be equal or slightly larger than the foundation/base slab.

3.5 BUILDING STAGES OF TWO SAMPLE OVENS

Although wood-fired ovens can be classified as commercial and residential/domestic ovens, they share the same basic structure. However, commercial ovens are typically larger in size and have a larger thermal mass due to thicker dome walls and more materials used. Commercial ovens are often built in a spheroid shape, which is a flattened sphere at the poles, resulting in a shorter height compared to the radius of the circular base. This design reduces heat loss and keeps the reflected radiated heat closer to the oven walls.

It is worth noting that commercial ovens require less firewood during the initial firing phase and have smaller temperature variations over time, reducing the need for frequent maintenance of the fire. On the other hand, smaller domestic ovens typically consume a significant amount of firewood throughout the entire firing and cooking cycle.

In this section, two oven designs will be outlined: free-standing structures with inner floor diameters of 120 cm and 160 cm. The construction stages of these ovens will be summarized by selected illustrations taken during the construction process, which are supported by explanations.

3.5.1 Basic Steps of a 120 cm Diameter Wood-Fired Oven Construction

This outdoor wood-fired oven is specifically designed and constructed under one side of a pergola, creating a convenient access point through an entertainment area that also serves as a shelter for the cook. The strategic location provides an easy visibility of the cooking process and the ability to hear the sounds of firing and cooking. This particular oven serves as a benchmark for temperature data logged in Chapter 5, showcasing its functionality and performance. Figure 3.10 provides a visual representation of the completed oven, which can be correlated with the construction stages (see Table 3.4) described in the subsequent figures (Figures 3.11–3.13).

TABLE 3.4   The descriptions of the building steps of the oven, see Figures 3.11–3.13

Step 1	A typical reinforced concrete foundation slab can be used as the starting point for building the wood-fired oven. The thickness of the slab and the strength of the wire mesh will depend on the size and weight of the final structure. In the case of a smaller oven, a regular mesh can be sufficient, and the foundation section can be filled with old wall bricks to minimize the amount of cement mixture needed. The foundation slab, measuring 1.6 m x 1.6 m, should be placed on well-drained soil without additional footings, with careful consideration given to the orientation of the oven facing the entertainment area.
Step 2	Once the concrete is poured, it needs to be accurately levelled using a spirit level and smoothed to facilitate the construction of the supporting walls for the raised slab. It is crucial to allow the slab to fully dry before proceeding with the construction of the supporting walls. Proper reinforcement of the base will ensure that the entire oven structure can be easily relocated if needed.
Step 3	After the foundation slab has dried completely, the support walls should be built and raised to a height of 0.85 cm. Special attention should be given to the door access at the bottom of the oven, which serves as a storage space for firewood. The top slab, when completed, will provide a final working height of 1 m.
Step 4	To accommodate the different levels of the cooking surface and the front working space/balcony, the raised slab area is divided into two sections by a divider, necessitating varying slab thickness. The use of a spirit level is critical during these construction phases to ensure accuracy.
Step 5	For the arch section of the oven door, curved bricks can be utilized, although a thick curved metal plate is preferred as a door arch due to its simplicity and structural rigidity to support the wall bricks above. The height of the wall will ultimately support the base slab.
Step 6	Before pouring the concrete, a construction frame needs to be built, along with reinforcing wire mesh aligned with the outer face of the walls and extended to the front section. In this specific design, an ash-slot construction frame is also incorporated in the middle of the front working space. It is important to ensure that the wire mesh is properly positioned in the middle of the intended slab thickness before pouring the concrete in one go. The construction phase of the raised slab should be completed and allowed to dry. Note that the front and back sections, where the dome will be constructed, are at different levels.

TABLE 3.4 *(Continued)*   The descriptions of the building steps of the oven, see Figures 3.11–3.13

Step 7	A custom-built stainless-steel metal flue kit is used for the chimney system, consisting of three standard-length pipes. The first flue pipe is attached to a custom-built metal door arch through welding. The construction of the oven door structure is completed using curved bricks for the front arch of the door cavity.
Step 8	An insulating thermal wool is laid over the slab. A bed of fine sand, approximately 5 cm thick, is then added on top of the insulating thermal wool. Once the sand is levelled, the centre of the oven floor is identified, and the thermocouple-tip installation is carried out by drilling halfway into a single paver. Two thermocouples are used and integrated into the oven structure: one in the floor brick and another at the top section of the dome. These thermocouples have long wiring (1.5 m) to reach the front wall where the signal conditioning and display unit is located. It is important to lay the floating floor bricks over the thermocouple wires.
Step 9	A typical reinforced concrete foundation slab can be used as the starting point for building the wood-fired oven. The thickness of the slab and the strength of the wire mesh will depend on the size and weight of the final structure. In the case of a smaller oven, a regular mesh can be sufficient, and the foundation section can be filled with old wall bricks to minimize the amount of cement mixture needed. The foundation slab, measuring 1.6 m x 1.6 m, should be placed on well-drained soil without additional footings, with careful consideration given to the orientation of the oven facing the entertainment area.
Step 10	The oven floor is constructed using high-density, high-temperature pavers, which are laid in a pattern that supports the floor's integrity. No mortar is used between the pavers, allowing them to move freely with changing temperatures and preventing cracking.
Step 11	The base of the oven is fully insulated at this stage, thanks to the raised slab, thermal blanket, fine sand, and floor bricks. The reference base circle, with a diameter of 120 cm, is marked on these bricks, considering the correct distance from the oven door and side walls.
Step 12	Careful sizing and shaping of the inner and outer arches for the door are crucial before their final assembly using mortar. The custom-built flue kit, including a door arch, aids in achieving this step. A front door arch, larger than the inner arch, along with a cardboard template placed in the inner arch, can be used to size the door. An ash-collection opening at the front workplace, connected to a drawer, is also noted.
Step 13	The front wall is built around the flue system, with a single curved brick indicating the starting position of the inner door arch construction. A recess between the outer and inner arches of the front door is necessary. The chimney recess is built around a customized metal structure welded to a flue piece.
Step 14	The dome construction, with a diameter of 120 cm, proceeds without using any formwork for the initial layers. The firebricks are placed together, with the edges of the inner face touching the neighbouring bricks without mortar. Mortar is only applied to the back of the bricks.
Step 15	The dome construction continues for a few layers, and a plywood formwork is built to match the inner dimension of the half-sphere, with a radius of 60 cm. Thick cardboard can be used to shape the top section of the formwork, providing an accurate guide and support for the firebricks.
Step 16	Once the dome is constructed and fully rendered with mortar, the insulating thermal blanket is placed. A wire mesh, such as chicken wire, is then applied over the thermal blanket, and a final coat of 3–4 cm thick mortar is added to complete the dome construction.
Step 17	For the flue system, a spark arrestor mesh is adapted inside the flue pipe, and a spark arrestor is cut and attached to the stainless-steel cowl. The flue kit, consisting of the flue pipe, spark arrestor, and stainless-steel cowl, is assembled. The height of the chimney should comply with building codes and be taller than nearby roof sections or other flammable materials.
Step 18	The final construction of the oven includes the roof frame, fully rendered outer walls, and tiling of the front workspace and door. The door is built to provide insulation, an airtight contact with the inner door arch, and handles necessary for slow cooking.

FIGURE 3.10   The finished view of the 120 cm domestic wood-fired oven that is enclosed by the brick walls and is covered with a red-tiled roof structure common in the Mediterranean region.

Ertuğrul, Nesimi, Personal photograph, "Final Dome-Shaped Oven," Adelaide, Australia, January 2015.

FIGURE 3.11 The construction steps showing the foundation and the raised walls, the top slab and the preparation of the floor base using the thermal wool and fine sand.

Ertuğrul, Nesimi, Personal photograph, "Dome-Shaped Oven Construction, Adelaide, Australia," November 2006.

FIGURE 3.12 The preparation of the floor base and the thermocouple connections, the construction steps of the floor and the door arches, and the beginning of the dome structure.

Ertuğrul, Nesimi, Personal photograph, "Dome-Shaped Oven Construction, Adelaide, Australia," November 2006.

FIGURE 3.13   The formwork for the dome, the final stages of dome insulation setting up the thermal blanket, setting up the wire mesh, the flue kit with spark arrestor and cap, the outside wall rendering, and the final door.

Ertuğrul, Nesimi, Personal photograph, "Dome-Shaped Oven Construction, Adelaide, Australia," November 2006.

3.5.2 Building Stages of a 160 cm Diameter Wood-Fired Oven, see Figure 3.14

FIGURE 3.14 Major construction steps of the 160 cm oven.

Ertuğrul, Nesimi, Personal photograph, "Dome-Shaped Oven Construction, Adelaide, Australia," January 2015.

REFERENCES

[1] Ramesh, M., Azenha, M., & Lourenço, P. B. Mechanical properties of lime–cement masonry mortars in their early ages. *Material Structure* 52, 13 (2019). 10.1617/s11527-019-1319-z

[2] Bompa, D. V. & Elghazouli, A. Y. Mechanical properties of hydraulic lime mortars and fired clay bricks subjected to dry-wet cycles. *Construction and Building Materials* 303, 124458 (2021), ISSN 0950-0618.

[3] Branco, F. G., Belgas, M. L., Mendes, C., Pereira, L., & Ortega, J. M. Mechanical performance of lime mortar coatings for rehabilitation of masonry elements in old and historical buildings. *Sustainability* 13, 3281 (2021). 10.3390/su13063281

[4] Jaine, T. *Building a Wood-Fired Oven for Bread and Pizza*. Prospect Books, 1996, ISBN 090732570X.

Ertuğrul, Nesimi, Personal photograph, "Tools and Utensils, Adelaide, Australia," 2008.

Cooking Utensils/Tools Firing and Fire Woods

4

4.1 INTRODUCTION

In order to ensure a successful wood-fired oven cooking experience, it is important to properly prepare the oven for cooking. This involves building a fire that will heat the oven to the desired temperature and maintaining the fire until the cooking is complete. The type of wood used for the fire can also have an impact on the flavour of the food, as different woods can impart different flavours.

When selecting wood for a wood-fired oven, it is important to choose a type of wood that burns cleanly and produces minimal smoke. Hardwoods such as oak, hickory, and maple are popular choices for their clean burn and pleasant flavour. Softwoods such as pine and cedar should be avoided, as they can produce excessive smoke and can also contain resin that can give food an unpleasant taste.

In terms of utensils and tools, traditional wooden paddles are commonly used to move food in and out of the oven, as well as to turn and remove pizzas and breads. Clay pots and cast-iron pans are also popular for cooking stews, soups, and other dishes in the oven. It is important to choose utensils and tools that are food-safe and can withstand high temperatures.

Having the right cooking utensils is crucial to fully utilize the benefits of the heat transfer characteristics of the wood-fired oven that were discussed previously. In addition, it is important to note that slow cooking at low temperatures for long periods can provide significant health benefits. This type of cooking requires utensils that can conduct and retain heat at a level necessary to achieve the desired outcome.

In the following sections, cooking utensils will be classified and explained based on their suitability for specific dishes. Pre- and post-cooking tools will also be discussed as they are necessary to ease cooking practices. The chapter will conclude with a discussion on firing the oven and fire maintenance practices, along with recommendations for the type of firewood to use in wood-fired ovens.

Prepare to embark on a culinary journey as the following sections delve into the essential cooking utensils, tools, firing techniques, and firewood selection required to unlock the full potential of wood-fired oven cooking. Embrace the rich flavours and captivating experience that wood-fired oven cooking brings forth.

4.2 COOKING UTENSILS

It is worth noting that while certain dough-based dishes like breads and pide can be cooked directly on the oven floor without the need for specific cookware, other cooking methods and various dishes may benefit from the use of appropriate utensils to achieve the desired characteristics of wood-fired oven

DOI: 10.1201/9781032640136-4

FIGURE 4.1 The types of cookware that can be used in the wood-fired ovens: fire place grits (a,b), cast iron grill pan (c), clay pot with lids (d), tajine (e), cast iron with lid (f), glass baking tray and glass pot with lid (g), shallow terracota cookwares and testi (potbelly shaped clay pot) (h), single-serve clay pots (i), sac (rounded outward stainless steel cookware) (j), stainless steel tray and skewers (k) and a horizontally rotating stainless-steel tool for doner/yiros (l).

Ertuğrul, Nesimi, Personal photograph, "Utensils, Adelaide, Australia," 2023.

cooking, as outlined in Chapter 2 and reported in Chapter 5. The choice of utensil plays a significant role in shaping the flavour of the final product, as it influences the heat transfer dynamics, including intensity and duration, between the wood fire heat source and the ingredients. Therefore, the selection of suitable utensils is pivotal in elevating the wood-fired oven cooking experience.

Figure 4.1 shows a range of cookware that is well suited for cooking in wood-fired ovens. These utensils cater to a wide variety of dishes that can be prepared using such ovens.

A fire grate (Figures 4.1a and 4.1b) is an excellent tool for grilling food directly over the wood fire during the fire reduction phase. The texture and crispiness of the food are influenced by factors such as cooking time, fat content, and fire intensity. For creating sear marks on meats while providing braising and searing capabilities, a pre-heated cast iron grill pan (Figure 4.1c) is a suitable option.

Clay-based cookware is closely associated with wood-fired ovens and comes in various shapes and sizes. These utensils are commonly used in Mediterranean, North African, and Asian cuisines. Figures 4.1d and 4.1e showcase two popular types of clay cookware: the clay pot (Romertopf, meaning Roman pot) and the tajine (or tagine). Earthenware with local names like Turkish guvecs, Greek giouvetsi, Spanish ollas and cazuelas, Japanese donabes, French poêlons, Chinese sand pots, and Moroccan tagines is also widely used. It is important for the pots to have a well-fitting lid.

Clay pots, with or without lids, are essential utensils for one-pot cooking in a wood-fired oven, closely followed by stainless steel trays. Clay is a thermally stable porous material with excellent adsorption properties, especially in the case of pot lids. However, the inner surface of the pot should not absorb liquids or react chemically with the food. Therefore, earthenware cookware typically features a glazed interior. Glazing is also crucial for preventing residual flavours and spices from previous cooking sessions, as unglazed pots can absorb and retain these essences within their pores.

The quality of the glaze in clay pots is vital for their durability and appearance. A high-quality glaze should have a smooth and consistent texture, free from cracks, bubbles, and other defects. It should also withstand high temperatures without chipping, flaking, or fading. Moreover, it is important to be aware of the potential health risks associated with specific glazes. Certain glazes may contain harmful chemicals or heavy metals that can leach into food during cooking or storage. Therefore, it is advisable to avoid glazes containing lead, cadmium, and barium, as they can pose serious health hazards. When purchasing clay pots, it is recommended to choose those that are certified as food-safe and made with high-quality, non-toxic glazes. Proper seasoning and cleaning of the pot prior to use are also crucial to remove any potential contaminants.

To prevent cracking, it is important to avoid sudden temperature changes before and after cooking with clay pots. It is recommended to soak the pots in water until thoroughly saturated before use.

After cooking, clay pots can be placed in the warm region (Region D as described in Chapter 5) of the oven or on a wooden board to minimize rapid changes in temperature. Thanks to their excellent heat retention properties, food cooked in clay pots will continue to cook even after being removed from the oven, making them an ideal choice for keeping dishes warm without extra effort. Furthermore, cooking in clay pots requires minimal fat, offering a healthier option for slow cooking, frying, baking, braising, and grilling. When making soup, for example, the dish simmers slowly and gently for extended periods without forming a skin, resulting in a highly flavourful outcome. Overall, clay pots provide an exceptional cooking experience for a wide range of slow-cooking practices.

The tajine, also known as a tagine, is a type of North African clay pot closely associated with Morocco, Tunisia, Algeria, and Libya. It features a large, shallow base and a conical lid, often with a hole at the top. The unique characteristic of the tajine, like other clay pots, is its ability to keep food moist and flavourful by allowing condensed steam to drip back into the stew during cooking. This cookware is perfect for slow-cooked stews with meat, poultry, or fish, providing a delightful one-pot meal for sharing. Additionally, the term "tajine" is used to refer to a Moroccan stew known for its blend of savoury and sweet flavours, often enhanced with ingredients like dates and prunes.

When selecting quality clay pots, it is important to choose pots that are glazed with a glass lining. This ensures food safety and makes clean-up easy. As mentioned earlier, several food safety criteria can be considered, such as FDA (Food and Drug Administration) testing and California Prop-65 checking, which provides a list of naturally occurring and synthetic chemicals known to cause cancer, birth defects, or other reproductive harm. DIN EN 1388-1 testing can also determine the potential leaching of lead and cadmium into food.

In addition to clay-based cookware, a variety of metal utensils can be used in wood-fired oven cooking, including aluminium, copper, stainless steel, cast iron, and carbon steel. The selection of metal utensils should be based on five key features: heat distribution, heat retention, reactivity, food safety, and cost. Both clay and metal cookware offer unique advantages and can be utilized to create delicious and flavourful meals in a wood-fired oven.

Figure 4.1 also showcases other cookware options. The cast iron cookware with a lid (Figure 4.1f) is renowned for its slow heat absorption, exceptional durability, uniform heat distribution, and excellent heat retention. The cast iron lid allows juices to continuously baste the food during cooking, making it suitable for searing, sautéing, simmering, braising, baking, roasting, and frying. While plain iron may leach into food, it is generally considered safe for cooking, especially when using enamel-coated cast iron.

A glass baking tray and glass pot with a lid (Figure 4.1g) are ideal for wood-fired oven cooking as they retain heat better than metal bakeware. Additionally, they do not have any adverse effects on the food, making them a safe and healthy choice for cooking.

Figure 4.1 also includes a selection of clay-based cookware in various sizes. Shallow terracotta cookware and Turkish testi (Figure 4.1h) as well as single-serve clay pots (Figure 4.1i) are smaller in size. The testi, a potbelly-shaped clay pot, is commonly used in the Cappadocia region of Türkiye for slow-cooking lamb, beef, or chicken with vegetables like carrots, onions, garlic, and potatoes. The dish itself is known as "Testi kebab" because of the vessel in which it is cooked. In practical cooking, the testi is sealed with bread dough and left to simmer in its own juices for several hours. Once the cooking is complete, the neck part is cracked open, and the food is served.

Figure 4.1j features a unique utensil known as a sac, which is a rounded outward stainless-steel cookware. An inverted wok can serve a similar purpose. However, the utensil needs to be preheated inside the oven before draping flatbread products over the convex part. This cooking style is commonly performed in Anatolia in the hearths of rural households to prepare larger and flat varieties of bread, such as "yufka," "katmer," and "gözleme."

In addition to the clay pots, another useful cookware option shown in Figure 4.1k is the stainless-steel tray. This cookware is often required for slow-cooking meats and roasting vegetables, such as aubergines and capsicums. Thin or round stainless-steel skewers that can rest on the edges of the tray can be used to make shish kebab varieties, which will be covered in Chapters 6 and 7. It is important to note that the maximum size of the stainless-steel tray depends on the width of the door cavity of the wood-fired oven.

When selecting cooking utensils for a wood-fired oven, two general criteria should be considered:

- The heat distribution and retention of the cookware play a crucial role in achieving effective and uniform heat distribution, as well as optimal heat retention based on its mass and material properties. As explained in previous chapters, cooking utensils made from clay or metals have different heat distribution and retention properties. Additionally, using a lid during cooking can offer faster, more uniform, and flavourful results.
- It is important to consider the reactivity of the cookware material with food and minimize any health risks associated with it. Utensils made of aluminium or copper are not recommended due to their reactivity. Coatings and linings should also be carefully assessed as they can wear down over time, potentially containing harmful or toxic materials. Lead-glazed pots, in particular, should be avoided. Paying attention to food safety is crucial, and if the clay quality is unknown or the purity of clay pots cannot be guaranteed, it is advisable to avoid using unglazed clay cookware.

Stainless steel utensils are a great option for wood-fired oven cooking as they do not rust or react with food, are easy to maintain, and are suitable for slow cooking and roasting. However, their heat distribution characteristics may not be as good as those of clay or cast iron cookware. Additionally, the thickness of the cookware plays a role in the duration of heat transfer, and it is best to avoid using a base metal plate for cooking flatbread, as it may result in uneven cooking.

It is important to note that conduction heating is the dominant heat transfer method in wood-fired ovens. When using clay cookware, it is essential to wash it immediately after use with hot water to minimize stains. Dishwashing substances should be avoided as they can seep into the porous clay and mix with the food during cooking. Clay cookware should be drained and air-dried or oven-dried at the lowest setting, and stored with the lid open. Careful handling of clay cookware is necessary due to its delicate nature.

The cost of utensils can vary significantly based on factors such as product quality, physical size, structural complexity, amount of material used, and operating temperature. When purchasing cookware, it is important to check the purity of the clay and the quality of the glaze. There are three main types of clay-based pot ranges available, each with unique characteristics to consider: earthenware, porcelain, and stoneware. The choice depends on individual preferences and cooking needs.

- Earthenware: contains impurities, is porous, opaque, and coarser in texture. It is typically fired between 600°C and 1,100°C.
- Porcelain: primary components are clays, feldspar or flint, and silica. It is characterized by a fine-grained, white body and is fired between 1,200°C and 1,400°C.
- Stoneware: a semi-vitreous ceramic made primarily from stoneware clay or non-refractory fire clay, fired between 1,100°C and 1,300°C. It is nonporous, usually glazed, and usually coloured grey or brownish due to impurities in the clay. It is impermeable and hard enough to resist scratching. Stoneware has been developed after earthenware and before porcelain and is commonly used for high-quality dishes. Chinese Ding ware is a typical example.

Additionally, it is worth noting that clay-based cookware should be washed immediately with hot water after use to minimize staining, and dishwashing substances should be avoided in porous clay pots as they can seep into the pores and mix with the food during cooking. Clay cookware should be drained and allowed to air dry on a kitchen towel or oven-dried at the lowest setting. Care should be taken when handling clay cookware as it is delicate. Finally, it is recommended to store clay cookware with the lid open.

In addition to the clay stew pan, there are many other traditional utensils that are perfect for cooking in a wood-fired oven. For example, the cazuela, a traditional Spanish dish, is often cooked in a wood-fired oven. It is a clay dish with a concave shape and glazed interior, making it ideal for cooking stews, roasts, and even paella. The donabe, a Japanese earthenware pot, is another option that is usually glazed on the inside and porous on the outside, retaining heat well. It is versatile, suitable for slow cooking, soups, stews, braised dishes, steaming, and roasting. In Chinese sand pot cooking, meat is blanched and/or caramelized first, then simmered slowly in the oven with vegetables and water until infused with a sweet-salty flavour. These traditional utensils are designed to work perfectly in a wood-fired oven, enabling home cooks and professional chefs to create dishes that are packed with flavour and texture.

4.3 PRE/POST-COOKING TOOLS

In addition to the essential tools used for preparing dishes and loading them into kitchen utensils, there are several specialized tools required for wood-fired oven cooking. These tools are listed in Table 4.1, accompanied by a brief explanation of their functionalities. While not all of these tools may be necessary for those who are new to wood-fired oven cooking, each of them serves a highly useful function that can enhance the cooking experience, making it more enjoyable and efficient.

TABLE 4.1 Wood-fired oven tools and accessories

NAME	TYPICAL IMAGE	FEATURES/REMARKS
Dough maker/mixer		A dough maker/mixer is a valuable tool for kneading, mixing, and blending ingredients when making dough. It is highly recommended as it simplifies the process of creating consistent dough. The mixer includes a mixing bowl, which is commonly made of stainless steel for its durability and easy maintenance. Additionally, it is equipped with kneading hooks that mimic the motion of kneading dough by hand. Key features of a dough maker/mixer include its capacity, mixing power, variable/multiple speed settings, timer functionality, and safety measures such as a bowl locking mechanism and finger guard. These features contribute to a seamless and efficient dough-making experience.
Brush to clean bread products		The bristle stiffness, size, and type of fibres used are key features of a bread cleaning brush, as they determine its effectiveness in cleaning various types of bread products. There are different types of bread cleaning brushes available, including bristle brushes, dough brushes, and crumb brushes. Some common features of these brushes include: • Material: Bread cleaning brushes are made from either natural or synthetic fibres such as nylon. • Cleaning Excess Flour: These brushes are suitable for removing excess flour from dough after kneading, ensuring a clean and neat finish. • Removing Debris: They are also useful for removing any debris from bread products after baking or after they have been sliced or cut, helping to maintain the cleanliness and presentation of the bread.
Dough scrubber (Bench scraper)		This is a versatile tool used for manipulating dough and other sticky substances during baking and cooking. It typically consists of a blade made of metal, plastic, or silicone, with a handle attached at one end. The blade is designed for various tasks such as scraping dough off surfaces, cutting or dividing dough into portions, and lifting and moving dough from one place to another. Its main purpose is to ensure efficient handling of dough while minimizing residue or waste.

TABLE 4.1 (Continued)   Wood-fired oven tools and accessories

NAME	TYPICAL IMAGE	FEATURES/REMARKS
Wooden spoons		They come in various sizes and shapes, making them versatile, durable, and eco-friendly utensils for a wide range of tasks. They do not scratch non-stick surfaces, are gentle on cookware, and do not conduct heat. However, they can absorb flavours and odours over time. Hence hand cleaning is recommended after each use to maintain their cleanliness and prevent the transfer of unwanted flavours.
Rolling pins		They are essential tools in baking and cooking, used to flatten and roll out dough or pastry, come as straight rolling pins, tapered rolling pins, textured rolling pins, and marble rolling pins, and can be made of wood, plastic, or metal. Turkish rolling pins, in particular, are long and thin without handles. They are slightly tapered at the ends and are used by pressing down and rolling the dough with the palms of the hands. These rolling pins are ideal for creating thin and even dough for various culinary creations. Note that a dough sheeter machine may also be used instead of a manual rolling pin.
Mitten and gloves		Both mittens and gloves for oven cooking are available in a variety of sizes and materials. They are commonly made from heat-resistant materials like silicone or fabrics such as Kevlar that can withstand high temperatures. These protective accessories are designed to cover the entire hand and wrist, ensuring maximum protection from hot surfaces and steam.
Heavy duty brass-bristle brush		It is a useful tool for maintaining a clean and well-functioning wood-fired oven. It should be used regularly when cooking pide and flatbread options to prevent the build-up of ash and debris that typically results from food pieces and flour. Ideally, it should have a long handle and a brush section with circular edges to reach the corners of the oven floor.
Wheel cutter		A wheel cutter or pizza cutter is an essential tool for making pide and pizza, as it is used to slice through pieces.

TABLE 4.1 (*Continued*)   Wood-fired oven tools and accessories

NAME	*TYPICAL IMAGE*	*FEATURES/REMARKS*
Ash shovel		An ash shovel is used to remove ash and debris from the fireplace in the oven. It is typically made of metal and has a long handle and a wide, flat rectangular or curved blade.
Peel paddles		Peel paddles are vital tools used for transferring pide, flatbreads, pizzas, and other utensils to and from the oven. Therefore, their size and rigidity are critical. They can be made of wood, metal, or a combination of both. Peel paddles come in various sizes and shapes, including circular, elliptical, and rectangular with round edges. The front or side working edges of the peels should be slightly chamfered to allow for easy sliding of the bread products. When handling dough products, a thin layer of flour should be used to prevent sticking. Note that the size, shape, and strength of the paddle can impact how much weight it can carry and how easy it is to manoeuvre. Wooden paddles are lightweight and can absorb moisture, which may cause warping over time. Metal paddles can be heavier, but they are durable, easy to clean, and can handle heavy utensils easily. The length of the handle can vary depending on the size of the paddle and its intended use. Longer handles can provide more reach for deeper ovens, while shorter handles may be easier to control for smaller ovens.
Basting (dump) mop and bucket		A basting mop is necessary at the beginning of cooking in the wood-fired oven to wipe down the inside of the oven after brushing and scraping to remove any remaining debris. It is also regularly used during cooking to remove food debris. It is important to note that during this cleaning phase, the floor temperature may drop by about 5–10°C.
Flour spoon		A flour spoon is used for measuring and scooping flour. It is a small, elongated spoon with a flat bottom and slightly curved sides, making it easy to level off the flour when measuring.
Pastry and pasta dough docker		A pastry and pasta dough docker is used for preparing flatbread and pide dough after it has been flattened. This handheld device features several rows of sharp pins or spikes, which are used to poke small holes into the flattened dough. Traditionally, this was done by fingertips.

TABLE 4.1 (Continued) Wood-fired oven tools and accessories

NAME	TYPICAL IMAGE	FEATURES/REMARKS
		Note that the pins or spikes of the docker penetrate the dough to create small, evenly spaced holes which prevent the dough from puffing up or forming air pockets or to store water layer that helps cooking various styles of bread products.
Dough ball tray		A proofing tray is designed for storing and proofing dough balls before they are shaped and baked. It is typically made of plastic or food-grade polypropylene and has multiple wells or compartments to hold individual dough balls. In Turkish cooking, a wooden tray with a proofing couche or wheat bran is traditionally used for proofing dough. The tray helps keep the dough balls separated and prevents them from sticking together, allowing them to proof or rise in a controlled environment, which enhances the texture and flavour of the final baked product.
Proofing couche cloth (Bread Flax Proving, Proofing Cloth)		A proofing couche, also known as a proofing cloth or a bakers' couche, is a special piece of fabric used in bread baking to support the dough during the final rise (proofing) before baking. It is usually made of linen or cotton with a dense weave that wicks moisture away from the surface of the dough, creating a dry, slightly crusty exterior. The ridges on the couche also help create an attractive pattern on the surface of the bread.
Infrared thermometer		A handheld infrared thermometer, also known as a non-contact thermometer or temperature gun, is a must-have tool. It can be used to measure the temperature of the oven floor without making direct contact. It is important to note that the accuracy of the temperature measurement is not critical ($\pm 1°C$), and the maximum operating temperature can be around 400–500°C. The emissivity value of the measured surface needs to be inputted into the settings to ensure accurate readings.
Pastry brush		It is a soft-bristled brush typically made from natural fibres like boar hair or synthetic fibres like silicone. This brush is highly useful for applying egg wash or melted butter to bread products before baking.

(Continued)

TABLE 4.1 *(Continued)* Wood-fired oven tools and accessories

NAME	TYPICAL IMAGE	FEATURES/REMARKS
Dough scoring tool (Bread lame, bread scoring knife)		A dough scoring tool, also known as a bread lame, is a tool used in bread baking to make shallow cuts or slashes in the dough before it is baked. These cuts help the bread expand and rise properly in the oven and give it an attractive appearance. The tool typically consists of a handle and a sharp replaceable or sharpenable blade. The blade is often curved or angled to allow for more control and precision when making cuts in the dough.
Pebbles, glass marbles or rock salt		These materials can be used in the wood-fired oven to create a different texture on the flat bread varieties.
Wooden lattices		A cooling rack can be used to cool down freshly cooked pide and bread products. It allows air to circulate around the baked goods, preventing them from becoming soggy on the bottom.
Moisture metre		A handheld humidity measuring device, also known as a moisture metre, can be used to measure the moisture content of firewood. There are two types: pin-type and pinless metres. When selecting a moisture metre for firewood, choose a model with a range suitable for the expected moisture content of the wood. Firewood typically needs to have a moisture content of less than 20% for efficient burning, so a range below 20% would be appropriate.

4.4 FIRING

As previously explained and illustrated in Chapter 3, it is recommended to locate the fireplace in a wood-fired oven on one side of the floor, separated by metal, brick, or accumulated ash barriers (if possible in larger ovens). This practice minimizes direct flame radiation, provides optimal heat transfer (conduction, convection, and radiation) for uniform cooking, creates a longer path for flames and hot air movement, increases heat retention, and improves energy efficiency. Additionally, this type of fireplace setup allows for easy fire setup and maintenance, while keeping the cooking surface cleaner.

To ignite the wood in the oven, dry kindling and paper are commonly used. Fire starters can also be used, although they are not necessary. Fire starters ignite instantly, producing high and long-lasting flames. They can be wrapped inside recycled or shredded papers within the fireplace. Kindling, which consists of small to medium-sized pieces of dry wood of any kind, is then placed over the fire starter. Pine and cedar woods are suitable choices for kindling.

4.4.1 Fire Starters

When selecting a fire starter, factors such as type, ease of use, and convenience should be considered. There are five main types of fire starters: ferro rods (ignited by friction, reaching temperatures up to 2,760°C), flint and steel (creating sparks to ignite a piece of cloth or tinder), magnesium rods (igniting while cutting, reaching temperatures up to 1,370°C), lighters (containing flammable liquid or compressed gas, creating an actual flame), and matches (creating an actual flame).

Considerations for selecting a fire starter may include size and convenience, ease of use, reliability in inclement weather, burn time, smell, and sustainability. Matches and lighters are easily accessible and low-cost options, along with kindling and papers. Commercial fire starters are also available, typically made from a combination of recycled wood fibres, paper, oil, and wax, and do not expire. Examples include wood/wax, pitchwood sawdust/paraffin wax, compressed oil in a pouch, wood chips/paper fibre/wax, pine wood/wool/stearin wax, and concentrated resin wood.

When selecting wood logs to use on top of the kindling, two key features to consider are burn time and heat output. Burn time refers to the total amount of time a single load of wood will burn, from ignition through to smouldering. Hardwoods such as oak, walnut, and cherry generally burn longer than softwoods like willow. It is important to use suitable firewood to avoid risks. Processed woods, woods with hazardous substances, laminated woods, treated and painted woods, or woods containing hazardous chemicals should be avoided. For example, wood with a high sap content like pine should be avoided, as the sap produces soot and creosote that can coat the oven floor and pose health risks.

Firewood should have low moisture content and be properly seasoned. High moisture content will result in shorter burns. The humidity in the convection heating and the type of wood used can impact the quality of Turkish pide. Willow tree branches, which have high moisture content, are used with established dry firewood inside the oven to adjust humidity levels and achieve a softer texture in the pide. Alternatively, mangrove tree branches or a stainless-steel tray filled with water can be used to control humidity in wood-fired ovens.

To achieve the optimum burn setting in a wood-fired oven, proper airflow and the correct size of the door cavity and flue are important. Low airflow fires burn at lower heat output for longer periods, while high airflow fires burn out quickly and require frequent addition of firewood. Leaving a thick bed of ashes in the fireplace helps insulate the fireplace and allows efficient heat transfer through the inner dome surface.

The installation of logs can affect the starting time and continuation of the fire, which is critical during the initial heating of the oven. Medium-sized logs can be added once the oven reaches the desired temperature level and the gasification phase has developed, maintaining a steady cooking temperature. Small kindling can be added later in the cooking process to provide light inside the oven for dough products such as pizza, pide, and lahmacun.

Factors such as altitude, humidity, and temperature of the air can have a minor impact on burn times. Ash build-up occurs over time in the fireplace, but it is not recommended to remove ashes until they overflow the heat barrier. In larger commercial-size ovens, the ash itself can be used as an effective heat barrier, resulting in highly uniform cooking for dough products.

The frequency and duration of firing in a wood-fired oven determine how often the flue pipe needs cleaning to ensure maximum airflow. If the flue pipe has a spark arrestor built-in, blockage can occur, depending on the quality and type of wood burned, necessitating regular cleaning.

4.4.2 Fire Woods

Wood fuels are widely used for heating purposes, and there are a variety of options available in the market. The main forms of wood fuels include firewood, wood chips, wood pellets, and wood briquettes. Firewood can be obtained in different forms, such as thin or thick, long or short-cut, and split oven-ready fuel wood. Wood chips are pieces of chipped woody biomass, while wood pellets are pulverized woody biomass, often in cylindrical form, with or without additives. Wood briquettes are produced by compressing pulverized biomass.

Hardwood is renowned for its density and weight, which allows it to provide a strong and long-lasting fire in wood-fired ovens. White and red oak trees are well known for their excellent firewood properties. Hickory trees, including pecan trees, also offer dense wood similar in quality to oak. Australian blackwood (*Acacia melanoxylon*) and red gum (*Eucalyptus blakelyi*) are other dense species suitable for firewood. Jarrah, a popular eucalyptus hardwood, is ideal for firewood due to its minimal smoke and ash residue.

White gum, a dense and heavy hardwood from the Eucalyptus family, burns slower and hotter. Sugar gum, although dense and heavy to handle, is considered a lower-grade firewood. River red gum has a low flame output and is well-suited for combustion wood heaters but may not be as effective in an open fire in a wood-fired oven. However, it is a commonly available firewood, along with red and grey ironbark.

In addition to Australian trees, white ash, sugar maple, and birch trees also yield excellent firewood. Sugar maple burns with minimal sparks and smoke. White ash is less dense and easy to split, while birch tree bark makes excellent kindling.

Softwood trees such as aspen, basswood, and willow are of poor quality for burning and generating heat. They provide less fuel and heat and are known to create creosote build-up when burned, which can adhere to chimney and flue walls.

When selecting the right type of tree for firewood, factors such as availability, hardness, cost, and specific cooking requirements should be considered. Generally, trees with needles instead of leaves should be avoided as firewood.

Oak wood produces a slow fire with glowing embers, making it an excellent choice for wood-fired ovens. Wood from fruit trees is also suitable for firewood, but it may not always be readily available. Once properly dried, fruit trees burn slowly.

In wood-fired ovens, it is important to use logs of suitable length (typically around 25–50 cm) and varying thickness to achieve a reasonable burn rate.

Mangroves are salt-tolerant evergreen forests found in tropical and subtropical regions along sheltered coastlines and shallow-water lagoons. Mangroves have high-density wood, and their charcoal burns hotter than most conventional charcoals. If available, branches and roots of mangroves are ideal for high-speed cooking and providing consistent heat in wood-fired ovens. Due to its unique aroma, it can add a subtle layer of nuance to cooked food.

In the western region of Turkey, olive pits are commonly used as fuel in wood-fired ovens. The residue, known as pirina (see Figure 4.2), is the raw seeds of olives obtained after olive oil production, which are then dried under the sun. Pirina is a highly combustible and cost-effective biomass fuel material. It offers uniform convection and conduction heat, making it ideal for bread cooking in the type of oven illustrated in Figure 3.2b in Chapter 3.

If the raw pirina has a high moisture and oil content, it is not advisable to use it directly as fuel. Raw pirina contains a significant amount of volatile matter, and gasifying volatile matters at low temperatures without complete combustion can have negative effects. Therefore, it is recommended to burn dry pirina with lower oil content after achieving high temperatures using a preheated fire. This increases the retention time of volatile matter at high temperatures. Additionally, the sulphur content in solid biofuels is generally lower than in fossil fuels, and sulphur tends to remain in the ashes (40–90%). While some

FIGURE 4.2   A pile of dry pirina ready for burning (left), pirina burning in the chamber (middle), and a pirina-heated oven showing flames through the heat transfer outlet (right). For more information on the structure of the oven, please refer to Figure 3.2b in Chapter 3.

Ertuğrul, Nesimi, Personal photograph, "Pirina and Pirina-Fired Oven, Turkey," May 2010.

publications claim it as a superior fuel source, the calorific value of pirina is around 18–21 MJ/kg, which is very close to the values of various types of wood.

In certain ovens, fruit residues obtained from processing various fruits (agro residue) are also used as fuels. In Argentina, a low-cost version of fruit pulp (such as apple and pear) called Biot is formed into log shapes after a sun-drying process (similar to cow dung used as fuel). Fruit pulp, rice husk, sawdust, and coconut shell powders have also been used with different binding agents (like dry cow dung, wheat flour, and paper pulp) to create briquettes under high pressure, ideally using environmentally friendly bonding materials. However, despite their higher densities, using briquettes may not be an economical solution in some cases compared to raw wood, as a piston-machine briquetting plant can have total factory costs ranging from 20–36 US$/ton of product.

Densities and heat values vary significantly among different wood species, and the volume of a stack of firewood can depend on whether it is split and how it is stacked.

Table 4.2 provides critical and comparative burning characteristics of some common wood species that can be used as firewood. These characteristics include a quantitative assessment of calorific value and a qualitative assessment of other features, such as ease of splitting, smoke, sparks, fragrance, and overall quality. Note that fragrance is also an important factor to consider when selecting firewood. For instance, poplar wood can emit an unpleasant odour when burned, while maple wood produces a sharp, pungent scent.

It is important to note that there are six main causes of smoky fires: wet wood or wood that contains resin, having a flue that is too small compared to the oven door, obstructions in the flue, downdrafts during windy periods, and a lack of oxygen. [1–7]

In addition, when deciding on the best option for heating, it is crucial to consider the heating capability and cost of different fuel types. Table 4.3 presents a comparison of the heating capabilities of various fuel types, including wood, electricity, and natural gas. This information can also be used to compare the cost of heating per unit. For instance, in 2023, the average cost of a cord (3.62 m^3) of oak wood ranges from US$180–300, while for maple wood, it is approximately US$350–450. On the other hand, the equivalent cost of electricity required to produce the same amount of heat varies from US$456–836 (based on residential electricity rates of US$0.12–0.22/kWh), and for natural gas, it is approximately US$230.

When selecting the best heating option, it is important to consider not only the cost but also factors such as availability, sustainability, and environmental impact of the fuel source.

TABLE 4.2   Characteristics of some commonly burned fire woods [8–15]

TYPES OF WOOD	CALORIFIC VALUE (MJ/KG)	EASE OF SPLITTING	SMOKE	SPARKS	FRAGRANCE	OVERALL QUALITY
Alder	17.5	Easy		Moderate	Slight	
Apple	27.0	Medium	Low	Few	Excellent	Excellent
Ash, Green	20.0	Easy	Low	Few	Slight	Excellent
Ash, White	24.2	Medium	Low	Few	Slight	Excellent
Aspen, Quaking	18.2	Easy		Few	Slight	
Basswood (Linden)	13.8	Easy	Medium	Few	Good	Fair
Birch	20.8	Medium	Medium	Few	Slight	Fair
Buckeye, Horsechestnut	13.8	Medium	Low	Few	Slight	Fair
Catalpa	16.4	Difficult	Medium	Few	Bad	Fair
Cherry	20.4	Easy	Low	Few	Excellent	Good
Chestnut	18.0				Good	Good
Maple, Other	25.5	Easy	Low	Few	Good	Excellent
Maple, Silver	19.0	Medium	Low	Few	Good	Fair
Mulberry	25.8	Easy	Medium	Many	Good	Excellent
Oak, Bur	26.2	Easy	Low	Few	Good	Excellent
Walnut, Black	22.2	Easy	Low	Few	Good	Excellent
Willow	17.6	Easy	Low	Few	Slight	Poor

TABLE 4.3   Heat values of various woods compared to other fuel types

3.62 M³ (1 CORD) OFAIR-DRIED WOOD	COAL (KG)	FUEL OIL (LITRE)	NATURAL GAS (THERM)	ELECTRICITY (KWH)
White oak, apple	900	480	174	3,800
Beech, sugar maple, red oak, yellow birch, white ash, black walnut	800	435	160	3,500
Grey and paper birch, black cherry, red maple, pitch pine	700	375	136	3,000
Black and green ash, sweet gum, silver and big leaf maple, red cedar, red pine	600	340	123	2,700
Poplar, cottonwood, black willow, aspen	500	284	102	2,200
White pine, white cedar	400	238	87	1,900

Selected energy levels of various fuel types:

Wood: 3.62 m^3 of wood and air (2.26 m^3 of solid wood at 20% moisture content). 1 kg = 14.44 MJ. Efficiency of the burning unit is 50%.

Coal: 1 kg = 29.12 MJ. Efficiency of the burning unit is 50%.

Fuel Oil: 1 lt = 38.52 MJ. Efficiency of the burning unit is 75%.

Natural Gas: 1 therm = 105.5 MJ Efficiency of burning is 75%.

Electricity: 1 kWh = 3.6 MJ. Efficiency is 100%.

In summary, when building a fire in a wood-fired oven, it is recommended to use dry hardwood as the main fuel source. Softwood can be used for kindling to ignite the fire, but it is important to split the logs for proper ignition. Adding some kindling or fire starters beneath the hardwood can aid in the ignition process. It is important to avoid using pine exclusively for firewood, as it can lead to the buildup of creosote in the flue or chimney. Hardwoods burn slower and longer with less resin, while softwoods burn more quickly.

The heat retention capability of a wood-fired oven depends on the materials used and the construction methods employed. Unlike electric or gas ovens, it can be challenging to accurately predict and measure the energy content of the primary fuel source, wood. Factors such as the type of wood (such as the root of the mangrove tree, poplar tree, cypress tree, red gum, etc.), the condition of the wood (old or new, moisture level), the size of the wood, and the heat level and timing of adding the wood all influence the amount of heat energy generated over a specific period of time. Therefore, measuring the thermal mass characteristics of a wood-fired oven and observing its firing experiences can provide valuable information about the heating properties of specific wood species.

4.4.3 Ash Removal and Disposal

Unlike wood-burning heaters, it is not necessary to remove the ash from wood-fired ovens after every cooking session. In fact, the ash can provide benefits by supporting the fire and insulating the bottom floor of the oven. It is recommended to keep the ash until it reaches the heat barrier level, which typically occurs after approximately 10–15 cooking sessions in a 120 cm diameter oven with a 12 cm high heat barrier. If the ash overflows the heat barrier, any excess ash can be safely removed. However, it is important to note that hot coals may remain hot and smoulder for several days, so it is advisable to consider this before immediately removing the ash after a cooking session.

It is worth mentioning that the ash residue can be beneficial as a substitute for lime in lawns and gardens, as it is rich in potash and can increase soil pH when applied. The recommended application rate for ash is generally around 5 kg per 10 m² of garden space per year, although specific recommendations may vary depending on soil composition and reaction. For instance, acidic soils with a pH less than 5.5 are likely to benefit from the addition of wood ash, while slightly acidic soils with a pH of 6.0 to 6.5 should not be negatively affected by ash application. However, wood ash should not be used on neutral or alkaline soils with a pH of 7.0 or greater. The acidity or alkalinity level of the soil can be measured using a pH instrument. It is important to consider the specific needs of plants when applying wood ash, as certain plants like asparagus and juniper are more tolerant of slightly alkaline conditions, while acid-loving plants such as potatoes, rhododendrons, and blueberries should not be exposed to wood ash.

REFERENCES

[1] Yun-Sang, C. et al. Comparative study on the effects of boiling, steaming, grilling, microwaving and superheated steaming on quality characteristics of marinated chicken steak. *Korean J. Food Sci. An* 36 (1), 1–7(2016). DOI 10.5851/kosfa.2016.36.1.1.

[2] pH Imbalance in the body: symptoms, causes, treatments, available at https://www.healthline.com/health/ph-imbalance#causes.

[3] Your guide to slow cooking, CSIRO Total Well being diet, available at https://blog.totalwellbeingdiet.com/au/2020/your-guide-to-slow-cooking/.

[4] Dumanoğlu, Y. & Bayram, A. Pirinanin Yakit Olarak Kullanimi: Stokerli Bir Kazanda Yakma Denemeleri, Yanma ve Hava Kirliliği Kontrolu VI. Ulusal Sempozyumu 10-12 Eylül 2003, İzmir, Turkey.

[5] The best fire starters of 2022, available at https://www.bobvila.com/articles/best-fire-starter/.

[6] Krajnc, N. *Wood fuels handbook.* Food and Agriculture Organization of the United Nations Pristina, 2015.

[7] Eriksson, S. & Prior, M. *The briquetting of agricultural wastes for fuel. No. 11.* Food and Agriculture Organization of the United Nations, 1990.

[8] Wood heating, available at https://forestry.usu.edu/forest-products/wood-heating.

[9] Burning wood and coal, by Susan Mackay, L., Baker, D., Bartok, J. W., Jr., and Lassoie, J. P. *Northeast Regional Agricultural Engineering Service, Riley Robb Hall.* Ithaca, NY: Cornell University, 14853. (607) 256-7654. 90 pp, 1985.

[10] The wood burner's encyclopedia, by Shelton, J. and Shapiro, A. B. 1976. Vermont Crossroads Press, Box 333, Waitsfield, VT 05673. 155 pp.

[11] Wood heat safety, by Shelton, J. 1979. Garden Way Publishing Co., Charlotte, VT 05445. 165 pp.

[12] Heating with wood and coal, revised by Bartok, J. W. 2003, Natural resource, agriculture, and engineering service (NRAES), ISBN 0-935817-91-3.

[13] Wood ash in the garden, available at https://www.purdue.edu/hla/sites/yardandgarden/wood-ash-in-the-garden/, accessed on 8/3/2022.

[14] Vivek, C. P., Rochak, P. V., Suresh, P. S., & Kiran, K. R. R. Comparison study on fuel briquettes made of eco-friendly materials for alternate source of energy. *IOP Conference Series Materials Science and Engineering* 577(1), 012183 (December 2019), DOI: 10.1088/1757-899X/577/1/012183.

[15] The art of wood fired cooking, by Mugnaini, A., Thess, J., and Smith, G. ISBN: 9781423606536.

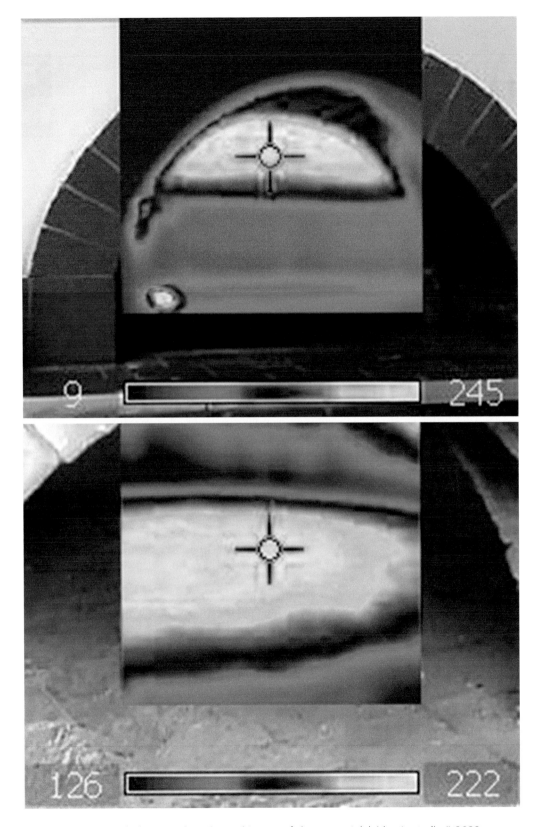

Ertuğrul, Nesimi, Personal photograph, "Thermal Image of the oven, Adelaide, Australia," 2022.

Data Logging and Analysis

5

5.1 INTRODUCTION

The data logging and analysis chapter of the wood-fired oven book investigates into the exploration of precise measurement techniques to improve cooking practices and enhance the understanding of a wood-fired oven's thermal behaviour. While traditional methods like burning flour on the oven floor can provide a rough indication of temperature, this chapter introduces more advanced methods that offer scientific insights into the wood-fired oven's performance.

Unlike technologically advanced ovens with precise heating controls, wood-fired ovens rely on their high thermal mass to provide stable temperature regions over extended periods, making them suitable for various cooking methods. However, gaining a deep understanding of a wood-fired oven's behaviour often requires extensive cooking experience. Therefore, this chapter focuses on two advanced measurement and diagnostic methods: data logging hardware with software and thermal imaging cameras.

By utilizing data logging and thermal imaging techniques, it becomes possible to capture key characteristic features of the oven under test and minimize the time needed to develop practical experiences regarding its thermal behaviour. These insights are essential for successful cooking and can contribute to achieving uniform results.

The chapter begins by offering definitions and explanations of the main measurement methods and their limitations. The intention is not to favour one specific method over others but rather to provide general conclusions and critical insights into the heat distribution within wood-fired ovens, particularly those with dome or igloo structures. These insights enable users to rapidly comprehend the thermal behaviour of their own ovens, which can be obtained after just a few firing and cooking sessions.

Throughout the chapter, the validity of achieving uniform cooking in wood-fired ovens, even with a rough reference temperature, will be demonstrated. The subsequent sections will verify these concepts and provide practical evidence to support them.

By performing data logging and analysis techniques, this chapter equips readers with the knowledge to better understand and harness the potential of their wood-fired ovens, ultimately enhancing their cooking experiences and achieving consistent results in a wide range of recipes.

5.2 DEFINITIONS

The utilization of thermal mass materials in ovens can effectively minimize temperature fluctuations. Similar to their application in houses, high thermal mass in ovens can yield comparable benefits, as

DOI: 10.1201/9781032640136-5

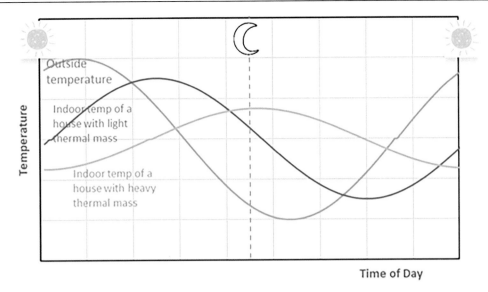

FIGURE 5.1   The impact of thermal mass about the heating characteristics of houses.

Ertuğrul, Nesimi, Personal Drawing.

shown in Figure 5.1. In this illustration, direct solar heating during the day contributes to free winter-time heating when the thermal mass of a house is exposed. This phenomenon holds true for double-brick houses with tile roofs, which share some similarities with wood-fired ovens possessing high thermal mass. The figure emphasizes the advantages of employing ovens with high thermal mass, which exhibit slower temperature changes over time. This characteristic proves advantageous when using instruments with slower response times to monitor temperature. As evident in Figure 5.1 for houses and later validated for ovens, thermal mass diminishes temperature fluctuations by storing and releasing heat. Consequently, wood-fired ovens with higher thermal mass render the outdoor temperature impact negligible during open-door cooking, irrespective of the season.

Figure 5.2 summarizes a variety of temperature measurement devices suitable for use in wood-fired ovens. For optimal results, it is recommended to employ an instrument that measures the temperature of the cooking surfaces, such as the floor and lower sections of the dome walls, prior to placing food in the oven. It is crucial to note that the analogue instrument depicted in Figure 5.2a can solely measure temperature at the specific point where it is affixed, limiting its precision and adaptability.

Figure 5.2 presents various temperature measurement instruments suitable for wood-fired oven cooking. One cost-effective option is the handheld infrared thermometer (depicted in Figure 5.2b), which offers non-contact temperature readings with a reasonable accuracy of up to 2% and a resolution of 1°C within a permissible range of up to 500°C. These devices provide flexibility to measure the temperature of specific locations on cooking surfaces and often come with a built-in laser pointer. However, caution should be exercised when using the laser beam function to prevent potential harm to the eyes, including reflected laser beams.

Thermal imaging cameras (shown in Figure 5.2c) detect temperature by capturing the invisible infrared radiation emitted by objects. These cameras consist of a lens, a thermal sensor, and processing electronics with an instantaneous display screen. When selecting a suitable thermal imaging camera, important specifications to consider include the range (the span of temperatures the camera can measure), field of view (the extent of the scene visible through the camera lens), resolution (the number of pixels on the scene), thermal sensitivity (the smallest detectable temperature difference), focus (manual or automatic), and spectral range (wavelength range detected by the camera sensor).

For accurate temperature measurements, the target being measured should completely fill the field of view of the thermal camera. It is worth noting that thermal cameras typically have lower resolution than

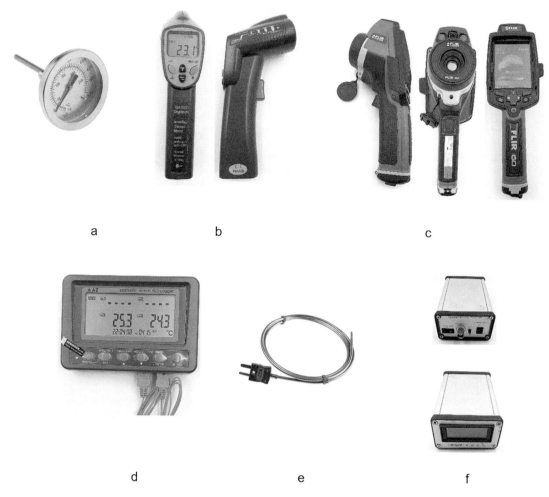

FIGURE 5.2   The methods of temperature measurements in the wood-fired ovens: a) An analogue thermometer, usually attached to the door panel or near the door. b) A hand-held non-contact infra-red thermometer with a laser pointer. c) A thermal imaging camera, non-contact. d) A 4-Channel temperature data logger with SD card and K-Type thermocouples. e) A K-Type thermocouple with a maximum temperature range up to 500°C, which can be integrated into the oven structure. f) A custom-built K-Type thermocouple amplifier with an indicator.

Ertuğrul, Nesimi, Personal photograph, "Measuring Instruments, Adelaide, Australia," 2022.

visible light cameras because thermal detectors require larger sensor elements to sense energy with larger wavelengths than visible light. Therefore, a thermal camera usually has fewer pixels compared to visible sensors of the same size.

In this book, the Flir E60 thermal imaging camera (depicted in Figure 5.2c) is used for temperature measurements. It has a temperature range of −20°C to +650°C, which covers the operating temperatures of the wood-fired oven. The camera boasts a high resolution of 76,800 pixels (320 × 240 × 240), a sensitivity of 0.05°C, and an accuracy of 2%.

It is important to note that emissivity, which measures an object's radiating efficiency, can vary based on surface condition, temperature, and wavelength of measurement. To obtain accurate temperature readings, emissivity serves as a modifying factor in single-colour thermometry. The emissivity value used in the thermal imaging camera settings should be an average of the individual emissivity

TABLE 5.1   Total emissivity of various surfaces which can be used in the wood-fired ovens

MATERIAL	EMISSIVITY
Fire brick	0.750
Earthenware	0.90
Concrete	0.94
Refractory white	0.90
Refractory black	0.94
Building brick (rough, no gross irregularities)	0.93
Copper calorized, oxidized	0.18
Glass fused quartz	0.75
Glass Pyrex	0.90
Cast iron oxidized	0.64–0.78
Iron cast polished	0.21
Stainless steel polished	0.16
Stainless Steel 303	0.74
Stainless Steel 304 (8Cri 18Ni) after heating	0.440–0.360
Wood, oaked	0.89

Note: When range of values for emissivity are given, linear interpolation can be done.

factors across the entire radiation spectrum being utilized. Table 5.1 provides emissivity values for selected materials found in the wood-fired oven structure.

To gather temperature variation data in wood-fired ovens, a suitable temperature data logger, as shown in Figure 5.2d, can be used. The data logger should possess the following desirable characteristics: an adequate number of channels to measure analogue temperature quantities, including ambient temperature; sufficient memory for data storage and easy data transfer; compatibility with the required type of thermocouples and temperature range (not exceeding 500°C in wood-fired ovens); date/time stamping capability; and programmability for sampling time. Ideally, the data logger should also be compact and portable. In this study, the Model 88698, a 4-Channel SD logger with K-Type thermocouples, was employed to capture data, which was further verified using K-Type thermocouples (Figure 5.2e).

To measure the temperature of specific locations or surfaces inside the wood-fired oven, K-Type thermocouples with a maximum temperature range of 500°C have been integrated into the oven structure as explained in the previous chapter. The thermocouples used have a stainless-steel mineral-insulated flexible probe sheath that was bent and shaped to suit the oven structure before being attached to the data logger. Alternatively, a custom-built K-Type thermocouple amplifier with an indicator can be used to provide a permanent display of the oven temperature, as shown and implemented in Figure 5.2f.

Figure 5.3 illustrates a traditional method that has been used in Türkiye for centuries to measure the temperature of wood-fired ovens, specifically to determine the correct temperature for cooking flatbread dough products, without relying on any instruments. This method involves sprinkling a small amount of flour over the floor bricks and observing the colour change of the flour after approximately 10 seconds. If the flour turns a darker brown, it indicates that the correct temperature has been reached. This method has been verified in the oven under test using the measurement methods shown in Figures 5.2b, 5.2c, 5.2d, 5.2e, and 5.2f. It has been determined that the ideal temperature range for flatbread dough products, including pizzas, is around 270–280°C.

There is a traditional method that has been used in Türkiye for centuries to measure the temperature of wood-fired ovens, specifically to determine the correct temperature for cooking flatbread dough products, without relying on any instruments. This method involves sprinkling a small amount of flour

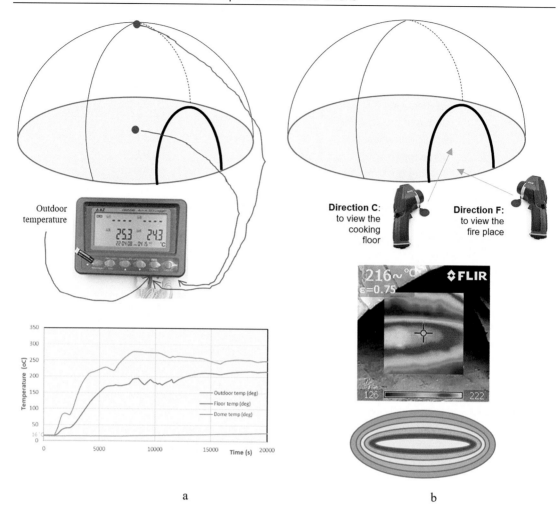

a   b

FIGURE 5.3 Two methods of measurements performed on the wood-fired oven under test. a) Data logging hardware connected to the two integrated K-Type thermocouples (red dots) and a third one for the outdoor temperature measurements. b) The thermal imaging can be used to capture the floor and the dome thermal images, from two different angles as illustrated.

Ertuğrul, Nesimi, Personal photograph/drawing, "Measuring Methods, Adelaide, Australia," 2023.

over the floor bricks of the fired oven and observing the colour change of the flour after approximately 10 seconds. If the flour turns a darker brown, it indicates that the correct temperature has been reached. This method has been verified in the oven under test using the measurement methods shown in Figures 5.2b, 5.2c, 5.2d, 5.2e, and 5.2f. It has been determined that the ideal temperature range for flatbread dough products, including pizzas, is around 270–280°C.

5.3 DATA LOGGING AND ANALYSIS

Before presenting the measured results obtained from the wood-fired oven, it is essential to familiarize ourselves with the two different methods of measurement employed. Figure 5.3 below serves as a guide,

TABLE 5.2 A comparison of the thermal camera and the thermocouple-based measurements

THERMAL CAMERA-BASED	THERMOCOUPLE-BASED
• *Non-contact and non-intrusive:* It can measure the temperature without physically touching and does not interfere with the cooking process. • *Comprehensive temperature profiling:* It captures temperature distributions across the entire oven surface, providing a visual representation of the heat patterns and hotspots within the oven. • *Real-time monitoring:* It enables continuous observation of temperature changes and heat flow during the cooking process. • *Larger area coverage:* Thermal cameras can capture temperature information from a wide field of view, allowing for the assessment of temperature gradients and variations across larger areas of the oven.	• *Point-specific measurements:* It offers direct temperature readings at the exact points of measurement. • *Intrusive/contact measurement:* It should have a physical contact with the oven surfaces, hence requires pre-installation. • *Higher accuracy:* When calibrated and use correct device, it can offer high accuracy measurement, which is not required in the wood-fired oven. • *Individual and flexible temperature monitoring:* If resources permit, thermocouples can be installed in a number of specific locations in the 3-D structure for simultaneous measurements.

illustrating how data was gathered using a data logging hardware (Figure 5.3a) and a thermal imaging camera (Figure 5.3b). In the subsequent subsections, the measured results will provide an in-depth analysis of the wood-fired oven structure.

Note that thermal camera-based measurements provide a broader view of temperature distribution and real-time monitoring, while thermocouple-based measurements offer point-specific temperature readings and individual monitoring capabilities. Although both methods have their advantages, they can be used complementarily to obtain a comprehensive understanding of temperature profiles in wood-fired ovens as implemented in this book. Table 5.2 compares these two complementary measurement methods.

5.3.1 Thermocouple Based Data Logging and Analysis

In the wood-fired oven, as mentioned previously and illustrated in Figure 5.3 the dome and floor temperature are measured using thermocouples that are integrated into the oven structure. These thermocouples are specifically designed to withstand the high temperatures of the oven environment. To monitor these temperature measurements, a 4-Channel SD logger was employed, which has the capability to connect and record data from multiple thermocouples simultaneously. Two channels of the logger are dedicated to monitoring the dome and floor temperatures in two strategic locations. Additionally, the third channel of the logger is utilized to measure the ambient temperature, which aims to assess the influence of the ambient temperature on the overall temperature performance of the wood-fired oven.

Figure 5.4 presents the temperature data collected throughout the complete cooking cycle of the wood-fired oven that underwent testing. The cycle lasted for a total of 29.8 hours and encompassed various phases, including the initial firing, open-door cooking with an active fire, closed-door cooking for approximately 11 hours after the fire extinguished, and a refiring period starting from an initial temperature of 110°C.

In Figure 5.4a, the temperature profile during the active firing phase is depicted. During this phase, the temperature rises on both the floor and dome surfaces, peaking at a maximum temperature of 277°C after approximately 3.5 hours. Subsequently, the temperature begins to decrease as the flame radiation subsides. It is common for wood-fired ovens to remain lit for a certain period to sustain the heat for subsequent cooking sessions. This phenomenon is evident in the temperature data, which exhibits small hill-like variations. The short, left-side of the uphill indicates active burning, while the longer, right-side downhill represents the reduction (gasification) period of wood burning.

(a)

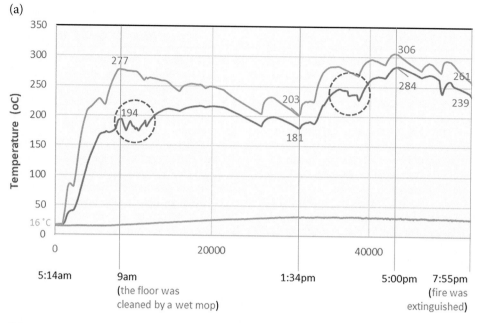

(b)

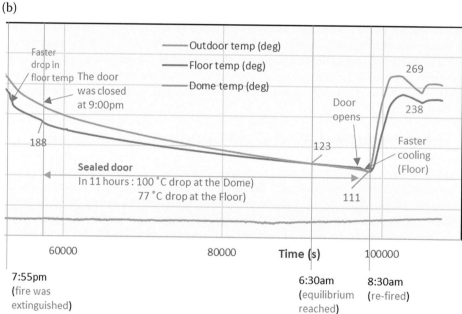

FIGURE 5.4 The temperature data recorded during the complete 29.8-hour cooking cycle. The data is divided into two sections for clarity: a) Section A (top) corresponds to the initial phases of the cooking cycle. It encompasses the firing process, heating stage, open-door cooking, and the subsequent extinguishing of the fire. b) Section B (bottom) represents the later stages of the cooking cycle, on the closed-door slow cooking phase and the subsequent re-firing of the oven.

Ertuğrul, Nesimi, Personal Drawings, "Measured Graph of Oven, Adelaide, Australia," 2022.

Note that the full cooking cycle has been divided into two distinct sections: Section A and Section B. This division allows for a clear visualization and analysis of the different phases and transitions observed throughout the entire cooking cycle.

In the figures, key temperature points are indicated, highlighting the temperature differences between the dome and floor sections of the oven. These temperature variations can be attributed to the varying thermal masses of the dome and floor structures during the active fire phase. Additionally, the fluctuations in dome and floor temperatures are influenced by the intensity of the active fire. As more firewood is added, the flame length increases, leading to greater radiated heat.

After 5 pm, the temperature difference between the dome and floor becomes minimal. This can be attributed to the diminishing flame radiation and the absence of additional firewood being added during this time. Furthermore, the dome and floor temperatures are less affected by heat losses within their structures and evaporation processes during cooking. Notably, the temperature lines of the floor display the effects of water-based cleaning, as indicated by the blue dashed circles.

It is important to avoid rapid temperature rises in the dome, as these indicate active burning with intense radiated flame heat transfer. Such conditions can also cause visible burns and result in uneven cooking of dough-based foods. Therefore, it is advisable to prevent these rapid temperature increases. Additionally, for searing meats, browning vegetables or casseroles, and general cooking purposes, the roasting temperature range of 230–300°C can be utilized. However, when cooking in this range, it is recommended to use a lid and maintain higher liquid levels to prevent over-browning or burning on the outer surface.

The consecutive figures from Figure 5.5 to Figure 5.8 provide further insight into specific aspects of the oven's behaviour. One key factor to consider is the concept of retained heat, which has the slowest time constant due to the absence of an active heat source. This absence makes the "no active heat source" temperature range ideal for slow cooking. Typically, this temperature starts around 150°C and gradually decreases based on the insulation quality of the entire oven structure, primarily influenced by the quality of the door panel and door seals.

The thermal time constant, also referred to as "thermal lagging," indicates the speed at which the temperature rises or decays in a given oven with a specific thermal mass. This time constant is an essential parameter to understand the oven's thermal behaviour. In this book, the rate of change of temperature ($\Delta T/\Delta t$) is used to analyse the key characteristics of the recorded data.

Looking specifically at Figure 5.6, it is evident that the oven heating process began with a consistently fed active fire, while the outdoor ambient temperature was recorded as 16°C. By calculating the rate of change of temperatures using the recorded data, it was determined that the dome temperature

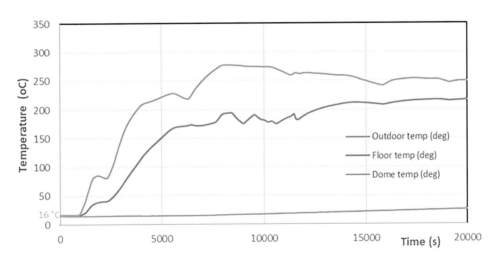

FIGURE 5.5   Outdoor, floor, and dome temperature variations during initial firing and heating the floor above 200°C.

Ertuğrul, Nesimi, Personal Drawings, "Measured Graph of Oven, Adelaide, Australia," 2022.

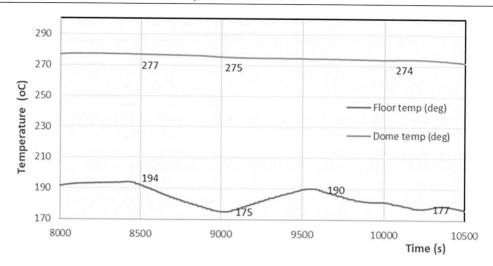

FIGURE 5.6 Temperature variations on the floor and the dome during the floor cleaning.

Ertuğrul, Nesimi, Personal Drawings, "Measured Graph of Oven, Adelaide, Australia," 2022.

exhibited a rate of 0.1°C/second (equivalent to 360°C/hour or 60°C/minute), while the floor temperature had a rate of 0.04°C/sec (equivalent to 144°C/hour or 24°C/minute).

These calculations provide valuable information about the heating rate of the oven and demonstrate the different rates of temperature change observed between the dome and floor. This analysis helps in understanding the dynamics of heat transfer and distribution within the oven during the heating phase as well as it quantifies and compares the thermal masses of the two distinct sections of the oven structure, the dome and the floor.

Figure 5.6, an enlarged section of Figure 5.5, provides a closer look at specific details regarding the oven floor and its cleaning process. It shows that the initial floor cleaning was performed using a wet mop at 9:00 am. The floor temperature variations observed indicate that cleaning actions were repeated multiple times.

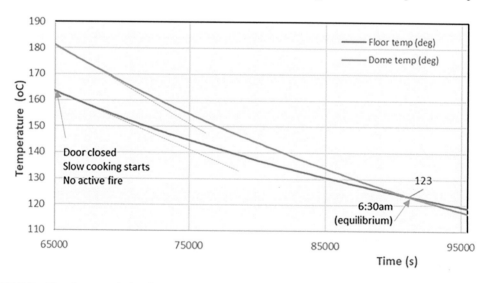

FIGURE 5.7 The dome and the floor temperature variations after the door panel is closed for the slow cooking.

Ertuğrul, Nesimi, Personal Drawings, "Measured Graph of Oven, Adelaide, Australia," 2022.

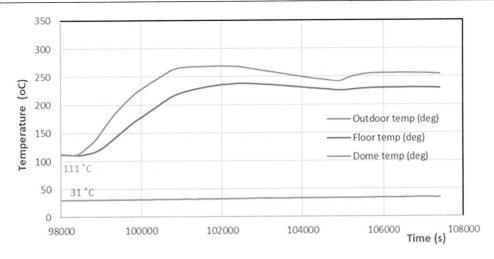

FIGURE 5.8   The dome and floor temperature variations after re-firing at 8:30 am.

Ertuğrul, Nesimi, Personal Drawings, "Measured Graph of Oven, Adelaide, Australia," 2022.

When the floor is exposed to cold water during cleaning, the temperature decays exponentially for approximately 8 minutes. Afterward, it begins to rise again and reaches a value close to the original temperature in about 8 minutes. It is worth noting that water-based cleaning has a more significant impact on the floor temperature compared to the dome temperature.

The small drop in both the floor and dome temperatures, typically around 3–4°C, is primarily due to the absence of flame radiation during the cleaning process. As the water evaporates, steam is produced, which extinguishes the active fire temporarily. However, the flame is reignited shortly thereafter, resulting in a swift recovery of the temperatures.

The amount of water used in the mop and the duration of cleaning can be adjusted based on the condition of the floor and the presence of any remaining food residues. This flexibility allows for effective cleaning while minimizing the impact on the oven's temperature profile.

By examining the temperature variations during the cleaning process, valuable insights can be gained regarding the dynamics of temperature change, the influence of water-based cleaning, and the subsequent recovery of heat in the oven.

After manually extinguishing the fire at approximately 7:55 pm (or waiting for the fire to naturally extinguish if a longer period was desired), two separate dishes were slow-cooked in the wood-fired oven during the test. The first dish consisted of wild goat meat placed in a stainless-steel tray and covered with aluminium foil. The second dish was a clay pot with a lid filled with various ingredients. The slow cooking process commenced when the door panel was tightly closed at 9 pm.

It is important to note that once the fire was extinguished, the floor temperature experienced a rapid drop due to the cold air sweeping across the surface. This phenomenon was previously observed in Figure 5.4b.

The entire slow cooking cycle lasted approximately 11 hours, and the temperature data for this phase is given in Figure 5.7. During this period, the rate of change of temperatures was calculated to be 0.0026°C/second (equivalent to 9.36°C/hour or 0.156°C/minute) for the dome and 0.0018°C/second (equivalent to 6.45°C/hour or 0.108°C/minute) for the floor. These rates of change indicate highly desirable values for effective cooking and are likely even better than those observed in many modern household ovens.

The exponential decrease in temperatures during this phase can be primarily attributed to heat losses in the dome and floor structures. These losses eventually reach an equilibrium point at approximately 123°C around 6:30 am (next day!). Following the equilibrium point, the temperature of the dome becomes lower than that of the floor due to greater heat losses. This discrepancy occurs because the dome is directly exposed to the open air, while the floor has been constructed over a thick raised slab, enhancing its thermal mass.

These observations provide valuable insights into the slow cooking process within the wood-fired oven, including the temperature profiles and the influence of different dish materials and cooking methods.

At the conclusion of the slow cooking cycle, the oven door was opened, and the oven was refired on the following day at 8:30 am, as illustrated in Figure 5.8. Notably, the initial starting temperature for this refiring was significantly higher at 110°C compared to the ambient temperature of 31°C (a hot summer day!).

During the refiring phase, the rate of change of temperature exhibited values similar to those observed during the initial firing. Specifically, the rate of change of temperature was calculated to be 0.07°C/second (equivalent to 252°C/hour) for the dome and 0.05°C/second (equivalent to 180°C/hour) for the floor.

However, it is important to acknowledge that the state and quality of the heat source, in this case, the firewood, will inherently differ in each heating attempt. These variations in the heat source can potentially influence the rate of change of temperature within the oven. It is crucial to consider the characteristics of the firewood, such as its moisture content and size, as they can impact the efficiency and heat output during the refiring process.

Understanding the rate of change of temperature during refiring allows for an assessment of the oven's performance and the effectiveness of the heating process. Monitoring and considering the quality of the heat source are essential factors to ensure consistent and optimal temperature control in subsequent heating attempts.

To showcase the sustained and prolonged temperature variation achievable in the wood-fired oven without the addition of firewood, Figure 5.9 has been presented. The figure demonstrates that within a timeframe of 1.4 hours, the temperature variation amounts to approximately 17°C, with the temperature ranging from 267°C to 284°C.

As previously mentioned, the thermal mass of the oven plays a crucial role in maintaining temperature stability. Higher thermal mass results in a more consistent and steadier oven temperature over an extended period. The thermal mass of the oven helps to absorb and retain heat, reducing temperature fluctuations and providing a more controlled cooking environment.

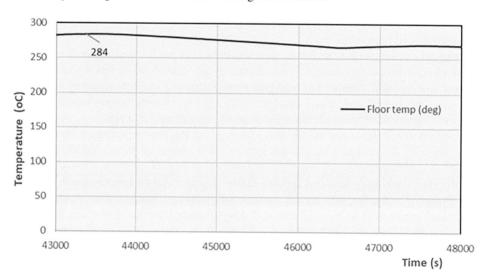

FIGURE 5.9   Demonstration of the steady temperature in the wood-fired oven.

Ertuğrul, Nesimi, Personal Drawings, "Measured Graph of Oven, Adelaide, Australia," 2022.

Figure 5.9 serves as evidence of the wood-fired oven's ability to sustain temperature variations without the need for continuous firewood replenishment. This feature is particularly advantageous for prolonged cooking processes that require a stable and controlled heat environment. The oven's thermal mass contributes to this temperature stability, ensuring consistent and desirable cooking results, which also demonstrate the capability of cooking wide range of recipes, outside the bread products.

5.3.2 Analysis of Floor Temperature Profiles for "Cooking"

In order to showcase the versatility of wood-fired ovens and their ability to cook a wide range of dishes beyond bread and pizza, Figure 5.10 presents the complete firing cycle of the floor temperature data, focusing on the cooking temperature ranges for a selected set of dishes.

It is important to note that while the temperature variation may appear significant, the scale of the figure can be misleading. The time intervals indicated between specific threshold levels allow for ample time to perform a variety of cooking tasks typically done in conventional ovens.

For instance, during the early phase of firing, roasting chestnuts, capsicums, or aubergines can be accomplished, or baklava can be cooked when the temperature ranges from 180 to 200°C. The baklava cooking relies on conduction and convection heat transfer, without the need for flame radiation, and usually takes around 20–30 minutes. Similarly, dishes like lahmacun, pide, and yufka bread require a temperature level of approximately 270°C, with a cooking time of only about 10 minutes.

It is crucial to emphasize that achieving a rise in oven temperature can only be accomplished by adding additional wood to the ongoing fire. Care must be taken to avoid excessive heat if aiming to cook at a specific temperature range for an extended period of time. Additionally, during open-door and active-fire cooking, it is possible to utilize clay pots with lids or trays covered with aluminium foil to contain the food. This approach can help mitigate the impact of flame radiation and result in shorter cooking times.

However, it is essential to recognize that nothing can truly replace the quality of "slow cooking" achieved through retained heat. This type of cooking, utilizing the temperature characteristics illustrated in Figure 5.10b, offers distinct advantages and can deliver exceptional results in terms of flavour and texture.

Overall, wood-fired ovens offer a wide range of cooking possibilities, allowing for the preparation of diverse dishes with varying temperature requirements. By understanding and leveraging the oven's temperature characteristics, one can harness the full potential of wood-fired cooking and create culinary masterpieces.

In general, convection and conduction heat transfer offer even heat distribution, which is preferred by some foods like bread and vegetables. However, certain temperature-sensitive foods such as soufflés, cakes, and muffins do not fare well with convection heat transfer. For these foods, it is recommended to cook them with the oven door closed, as given in Figure 5.10b. This closed-door cooking mode is also highly recommended for other dishes, including soups, stocks (made from bones for thicker liquid), and broths (a mixture of meat and vegetables for thinner liquid) that require gentle heat and extended cooking times. Additionally, toward the end of the slow cooking process (below 100°C), utilizing the oven's residual heat to dry fruits, vegetables, and herbs is recommended.

One significant benefit of slow cooking in a wood-fired oven is the extended cooling down period obtained during closed-door cooking. This allows for sequential or simultaneous cooking of different dishes, enhancing the versatility of this method. The wood-fired oven's ability to maintain a steady temperature range over an extended period proves particularly useful when cooking large cuts of meat or preparing slow-cooked stews and casseroles.

Another advantage of slow cooking in a wood-fired oven is the naturally low temperature levels before and after opening the door, which can be utilized for gently drying fruits like apricots, apples, figs, and tomatoes. This method of drying produces excellent results, as the heat is gentle and even, allowing the fruits to dry evenly without the risk of burning.

(a)

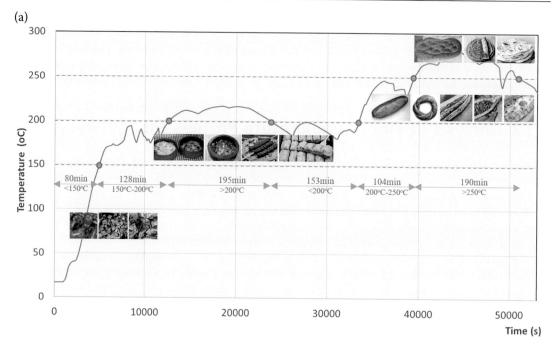

(b)

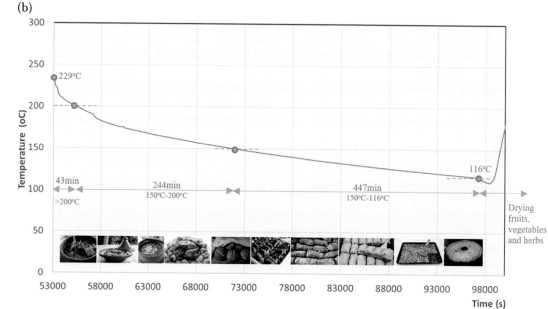

FIGURE 5.10   Full temperature cycle of the floor (29.8 hours in total) and a brief guide to the types of cooking within the temperature ranges highlighted. a) Open-door cooking region: Fire is started and oven is heated by adding fire wood, suitable for a range of breads, pides, lahmacun, baklava, meatballs, stuffed vegetables, casseroles, roasting, etc. b) Slow cooking region: Fire is extinguished and the door is closed, suitable for a range of dishes as illustrated and mentioned in the text.

Ertuğrul, Nesimi, Personal Drawings, "Measured Graph of Oven, Adelaide, Australia," 2022.

In conclusion, wood-fired ovens offer a unique and versatile cooking experience that goes beyond the traditional bread and pizza dishes. With their ability to maintain a steady temperature range for extended periods, wood-fired ovens are ideal for slow cooking, baking, and drying foods, making them a valuable addition to households or commercial restaurants.

5.3.3 Thermal Imaging Camera Data and Analysis

As shown in Figure 5.3b, a thermal imaging camera screenshot provides three essential pieces of data: the emissivity value set for the material being monitored ($\varepsilon = 0.75$), the spot temperature at the cursor's marked point (216°C, at the centre of the floor), and the temperature colour bar at the bottom of the screen, which automatically adjusts after defining the temperature range in the settings menu. Therefore, it is important to highlight that thermal images are presented with a coloured bar that represents the temperature range selected. As the colours progress from violet, blue, green, yellow, orange, red light, and white, the temperature increases.

For example, referring to the thermal image in Figure 5.3b and the accompanying explanatory drawing (oval rainbows), the oven's floor temperature is approximately 222°C (white) in the middle, while the entrance of the door and the bottom section of the dome measure around 126°C (blue). As the dome temperature rises, it reaches 222°C.

Images associated with "reference photos" indicating the fire starting and development phases are shown in Figure 5.11 while Figure 5.12 displays a selected sequence of thermal images that provide insight into the temperature pattern during the heating stage of the wood-fired oven.

FIGURE 5.11   Photos showing the fire starting and development phases utilized during the thermal camera tests.

Ertuğrul, Nesimi, Personal Photograph, "Thermal Images from the Oven, Adelaide, Australia," 2023.

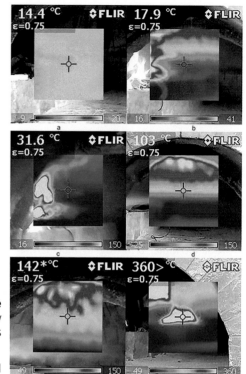

FIGURE 5.12 The thermal camera images as the **fire starting** and developing. The images are taken from the view of **Direction F** (see Figure 5.4) while the fire is developing as in Figure 5.11.

Ertuğrul, Nesimi, Personal Photograph, "Thermal Images from the Oven, Adelaide, Australia," 2023.

Note that the coating of soot on the side section of the dome in Figure 5.12 may suggest an area that is not yet adequately heated. However, this interpretation may not be accurate until the entire heating process is completed and sufficient heat energy is stored in the mass of the dome and the floor, which will be demonstrated in the subsequent thermal camera images. Additionally, the presence of soot on the dome depends on factors such as the length of the flame (as seen in Figure 5.12b), the condition of the firewood used, and the duration of the heating process.

Note that the thermal camera images were captured from two different directions, as indicated in Figure 5.3 (Direction C and Direction F), in order to observe the fire's development and the resulting temperature profiles at the top of the dome, side walls, and cooking floor. Figure 5.13 displays the changes in the top of the dome temperature during the wood combustion process, followed by the temperature profile in the dome's wall depicted in Figure 5.14. Finally, Figure 5.15 presents thermal camera images of the floor after it reached higher temperatures, indicating its readiness for cooking.

Based on the thermal imaging camera observations, several conclusions can be drawn:

- The top of the oven, being the highest point of the dome, experiences higher temperatures compared to the bottom floor due to two factors: the presence of long flames from the heat source (burning food) and the rising of hot air.
- Temperature variations around the floor are not uniformly distributed, which also supports the concept of separating the fire area from the cooking area.
- A smaller temperature difference between the bottom section of the walls and the nearby floor indicates more balanced conduction and radiated heat transfer to the cooking items.
- As the active fire subsides and the oven becomes ready for slow cooking, the temperature across the entire inner surface of the oven reaches equilibrium, creating ideal conditions for uniform heat distribution through natural convection.
- High-temperature white spots visible in the camera images indicate the presence of firewood pieces on the floor.

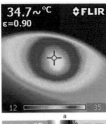

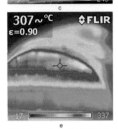

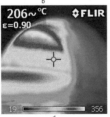

FIGURE 5.13 The changes in the **top** of the dome temperature during the combustion process of the wood. The camera images are obtained from the direct view, **Direction F** (see Figure 5.4).

Ertuğrul, Nesimi, Personal Photograph, "Thermal Images from the Oven, Adelaide, Australia," 2023.

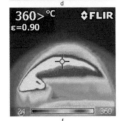

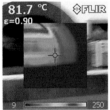
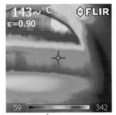

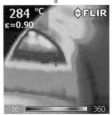

FIGURE 5.14 Changes in the **wall** of the dome temperature during the combustion process of the wood. The camera images are obtained from the side view of **Direction C** (see Figure 5.3).

Ertuğrul, Nesimi, Personal Photograph, "Thermal Images from the Oven, Adelaide, Australia," 2023.

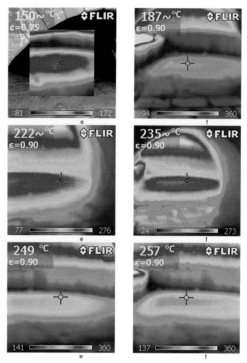

FIGURE 5.15 Temperature changes on the **floor** during the combustion process of the wood. The camera images are obtained from the view of **Direction C** (see Figure 5.3).

Ertuğrul, Nesimi, Personal Photograph, "Thermal Images from the Oven, Adelaide, Australia," 2023.

- The readiness of the oven for cooking flatbread products can be observed when the floor temperature reaches approximately 270°C, as seen in Figure 5.16f. It is important to note that at this point, the oven dome itself should be clear with no visible black soot. Since the active fire is separated on one side, cooking can commence immediately.

After conducting a comprehensive analysis of the thermal imaging camera data and verifying the results through real cooking experiments, several practical and valuable conclusions can be drawn. These findings are summarized in Figure 5.16., where different "Areas" (from A to H) represent distinct regions identified with relatively equal temperatures, indicated by different colours. The conclusions are as follows:

Area A: This designated active fire area is separated by a fire barrier and is suitable for cooking potatoes and corns wrapped in aluminium foil. This method produces softer, steamed end-products with enhanced flavour.

Area B: Due to its proximity to the active fire, cooking bread products in this area should be avoided. However, it can be utilized for placing clay pots with lids or other utensils covered by lids or aluminium foil. Careful attendance is required for fast roasting.

Area C: Similar to Area B, this region has less heat exposure and should be used with caution for certain cooking purposes.

Areas D and E: Area E, supported by the radiated heat from Area D, is the optimal cooking surface for bread products. Regular monitoring of the temperatures in these areas using a hand-held non-contact infrared thermometer with a laser pointer is recommended.

Area F: This region is ideal for keeping cooked items warm without further active cooking.

Areas G and H: Area G, along with Area H, is suitable for slow cooking with or without the door closed. When the door is closed, both areas reach equilibrium slowly, with conduction and radiated heat transfer being the primary modes. If the door is not closed, air circulation and heat losses may cause the temperature to drop quickly and unevenly.

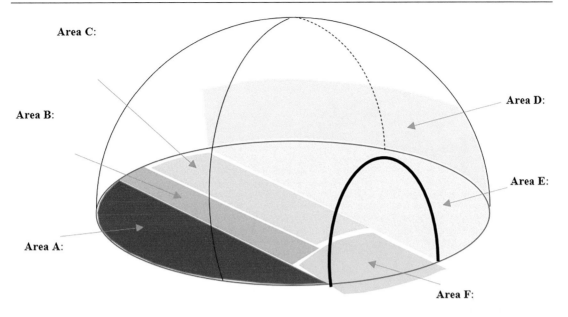

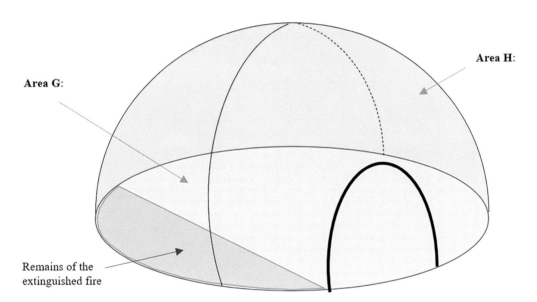

FIGURE 5.16   A summary of drawing highlighting the ideal cooking spaces in the wood-fired oven: Regular cooking phase with active fire source (top) and the slow cooking mode, no active fire (bottom).

Ertuğrul, Nesimi, Personal Rendering, "Thermal Images from the Oven, Adelaide, Australia," 2023.

As a general conclusion, it can be summarized that the active fire area can also be used for cooking. In addition, certain areas near the active fire should be used with caution. Other areas closer to the active fire have less heat exposure and should be used carefully for specific cooking purposes. Moreover, there are optimal cooking surfaces for bread products, supported by radiated heat from neighbouring areas. Furthermore, there is a designated region in the wood-fired oven ideal for keeping cooked items warm without further active cooking. Finally, certain areas are suitable for slow cooking, regardless of whether the oven door is closed or open.

Ertuğrul, Nesimi, Personal photograph," Shepperd Grandma, Mugla, Turkey," May 2010.

Turkish Cuisines

6

Classification of Common Features

6.1 INTRODUCTION

Turkish cuisine has a rich history that goes back thousands of years. Its uniqueness, which is characterized by the simplicity of meat and bread cooking and the freshness of the ingredients, can be traced to the historical lifestyle of the Eurasian nomads. However, Turkish cuisine has also been influenced by many other food cultures during the long migration routes that began in Central Asia. Despite some eating habits that have continued with minor variations throughout history, primarily based on ingredient availability, the cuisine has been very dynamic in accommodating new vegetables, seafood, and a range of herbs and spices due to cross-cultural influences along the migration routes.

This chapter explores the rich culinary heritage and diverse flavours of Turkish cuisine. As Turkic-origin people established empires and made permanent settlements, their eating habits were heavily influenced by local cuisines. This can be observed in the different culinary cultures, ethnic groups, and national dishes that have similarities in ingredients, taste, texture, and preparation method. Additionally, Anatolia (modern-day Türkiye) has a diverse range of regional variants that distinguish its cuisine. Many countries with Turkic-origin populations also have dishes with similar names, all cooked in various forms of wood-fired ovens, as mentioned earlier in the book. This dynamic nature of Turkish cuisine and its ongoing evolution is confirmed by its prevalence in many countries with Turkic minority populations, including Mongolia, Azerbaijan, Turkmenistan, Uzbekistan, Kazakhstan, the Kyrgyz Republic, Uyghurs, Georgia, Iran, Egypt, the entire Middle East, Afghanistan, Algeria, Libya, Tunisia, and even Europe.

In addition, it is worth noting that the long tradition of Turkish food culture is evident in the unique "single-word" names given to dishes. In contrast, western cuisine and culturally young dishes tend to have significantly longer names that list almost all the ingredients used. Some of these Turkish dish names include pide, iskender, karniyarik, keskes, dolma, sarma, baklava, kunefe, menemen, cacik, kokorec, kofte, and kebab. These well-established names directly relate to specific ingredients, with minor local variations, and describe the style of cooking and presentation. For example, if "doner" (gyros in Greek) is served over a plain pide flatbread, with a couple of tablespoons of yoghurt, decorated with grilled long green/red sweet/hot peppers, and finally served with freshly melted butter with or without tomato sauce, it is called "Iskender!"

DOI: 10.1201/9781032640136-6

While the lack of written documents makes it difficult to pinpoint the origin of many dishes, it is relatively easy to observe that current Turkish cuisine is largely the heritage of Eurasian nomads, Seljuks, and Ottomans. Therefore, Table 6.1 aims to provide an insight into this heritage, where basic meat, bread products, herbs, and spices have been utilized for over 5 millennia. For example, during the Ottoman period, many other cuisines, including Middle Eastern, Fertile Crescent, Mesopotamian, Near Eastern, Western Asian, North African, and French, had an influence on the Ottoman Palace kitchen, and vice versa. Additionally, cross-cultural influences are visible among Turkish, Greek, and Italian cuisines. Therefore, in this century, the culinary richness of Turkish cuisine, including a wide range of traditional foods, is at its peak and delightfully flavoured by the influence of the Silk Road on spice trades.

TABLE 6.1 Comparison of major ingredients of Turkish cuisine over the history

	HISTORY BY PERIODS AND REMARKS			
TYPES OF FOOD	*CENTRAL ASIA 3000BC- …*	*SELJUK AND BEYLIK ERA (1100–1300)*	*OTTOMAN (1326–1923)*	*MODERN DAY TÜRKIYE (1923- …)*
Meat and meat products	Lamb, horse, rabit, dear, conserved meat, pastirma,* sucuk**	Lamb, goat, horse, game birds, chicken, duck, turkey and few fish species, and sakatat, *** pastirma, sucuk	Lamb, lamb liver, beef and veal, chicken, turkey, duck, peacock and partridge, pastirma, sucuk, fresh/dry fish and various other seafoods: such as freshwater cod fish, eel, caviar, pickled bonito (skipjack tuna or toric), kipperfish, mackerel, mussels, oyster.	All three periods combined (except peacock) and remaining sea species, ox tongue
Bread products	Round flat bread types (pide, bazlama, yufka)	In addition to the previous time period, arı girde, çukmin, şebit and katmer.	White bread, oat bread, red corn bread, Nan-ı aziz and nan-ı hass (average quality bread), nan-ı adi (ordinary bread), fodula (white/flat bread), fırancala (white long loaf), yufka (large/thin flat bread) and somun (round bread). Highly influential French kitchen to the Ottoman Palace kitchen.	All three periods combined, and more dough products were created and influenced by other cultures (such as Italia, France and Middle East).
Culinary herbs, spices and likes		Indian knotgrass (madımak), edible spring weed (yemlik) and pennyroyal mint (yarpuz)	Spices: Black pepper, ambergris (amber) ****, mugwort (misk), chilli powder, vanilla allspice (yenibahar), cumin, saffron, mustard, coriander	Including all, combined with other herbs and spices as in Figures 6.10 and 6.11.

TABLE 6.1 (*Continued*)   Comparison of major ingredients of Turkish cuisine over the history

| | HISTORY BY PERIODS AND REMARKS | | | |
TYPES OF FOOD	CENTRAL ASIA 3000BC- …	SELJUK AND BEYLIK ERA (1100–1300)	OTTOMAN (1326–1923)	MODERN DAY TÜRKIYE (1923- …)
			seed and cinnamon. Herbs: Mint, parsley and basil	

Notes

* **Pastirma**: The word "pastirma" is derived from the Turkic noun "bastırma," meaning "pressing." It is a highly seasoned, cured, and air-dried beef that has its roots in Central Asia, brought by nomadic Turkic tribes and embraced in Turkish cuisine. To prepare pastirma, a cut of beef, typically from the hindquarters, is rubbed with a mixture of spices like garlic, fenugreek, and paprika. The meat is then cured in a cool, dry place for several weeks. The result is a firm, dark red meat with a robust, salty flavour. Pastirma is commonly thinly sliced and served as an appetizer or used as an ingredient in dishes like omelettes and legume casseroles.

** **Sucuk** (also spelled "soujouk" or "sujuk") is a type of Turkish sausage made primarily from beef. It is seasoned with various spices, notably black pepper, cayenne pepper, garlic, cumin, and sumac. Sucuk is a popular ingredient in Turkish cuisine, adding rich flavours to dishes like scrambled eggs, soups, and casseroles. It is often enjoyed as a breakfast food alongside eggs and bread. Sucuk is typically cured and dried, giving it a firm texture and allowing for long-term storage.

*** **Sakatat** (Offal meat): The term "sakatat" in Turkish refers to offal, which includes the internal organs and other parts of an animal that are not considered as "meat" cuts. In Turkish cuisine, sakatat is commonly used in a variety of dishes such as stews, soups, and grilled preparations. Examples of sakatat include liver (ciğer), sweetbreads (boyun), kidney (böbrek), brain (beyin), heart (kalp), tripe (işkembe), tongue (dil), head (kelle), intestines, feet (paca), and even testes. These parts are often cooked with onions, garlic, tomatoes, and a variety of herbs and spices to create flavourful and hearty dishes.

******Amber** (ambergris): Amber is a solid waxy substance that originates in the intestines of sperm whales. It is primarily used in Eastern cultures for medicinal and aromatic purposes, often found in perfumes and incense. In Turkish cuisine, however, it is not a common ingredient and is not typically used as a spice in culinary preparations.

In the following sections of this chapter, after covering the grains and breads, the ingredients and style of cooking are classified to highlight the commonalities and distinct features of Turkish cuisine before presenting few selected recipes in the final chapter. This chapter also aims to provide insights for future modifications and fusion cooking that can offer highly distinct tastes combined with the unique flavour of wood-fired oven cooking.

Note that the final chapter presents a carefully selected set of recipes to showcase the diversity of cooking in the wood-fired oven. These recipes have been chosen to highlight the unique flavours and techniques associated with cooking methods. However, for a broader range of recipes, including additional options and variations, a rich collection can be found in the link provided in Chapter 7. This link will lead you to a comprehensive resource where you can explore a wide array of recipes to further expand your culinary repertoire in wood-fired oven cooking.

6.2 GRAIN TYPES AND BREAD MAKING

Table 6.2 presents a compilation of whole grains commonly utilized in dough products cooked in wood-fired ovens. These grains can also undergo refining and processing steps to produce refined grains, which boast a longer shelf life compared to their whole-grain counterparts. This is because the oily germ present in grains can develop an unpleasant taste or odour when exposed to light, heat, air, staleness, or mould.

In Turkish cuisine, the most frequently refined whole grains include wheat, corn, rice, rye, oat, and barley. These grains are transformed into flour and directly incorporated into the dough of specific

TABLE 6.2 Major grains and their characteristic features

NAME	REMARKS	IMAGE
Amaranth	It is in pseudo-cereal group and has been cultivated for thousands of years as it was a staple crop for the Aztecs. It contains higher level of protein and has a peppery taste. Its flour is used in pancakes, biscuits, flat breads, pastas, salads, and baked goods, sprouted amaranth used in salads or cereals.	
Barley	One of the first grains to be widely cultivated around 8000 BC and originally native to Asia. It has a nutty flavour and chewy texture that works well in soups, stews, salads, and pilafs. Barley is also used to make flour for bread and other baked goods, as well as for breakfast cereals like oatmeal.	
Buckwheat	It is not a type of wheat, but rather a fruit seed. It has a distinctive earthy, nutty flavour. Buckwheat can be used in a variety of ways, including as a flour for making pancakes, crepes, and noodles (after mixing with wheat flour), as well as in bread, muffins, and other baked good, and can be used as a thickener for soups, stews, sauces, and casseroles, and is particularly popular in Eastern Europe and Asia.	
Corn	It can be consumed as a whole grain, ground into flour for making tortillas, bread, polenta and other baked goods, or used to make breakfast cereals. Corn is a versatile and important crop that has been cultivated for thousands of years and continues to play a significant role in global food production.	
Einkorn	Einkorn is an ancient variety of wheat that has been cultivated for over 10,000 years. It is a small-grained, hulled wheat and has a distinct nutty flavour and can be used in a variety of culinary applications. It is often used to make whole grain flour for bread, pasta, and other baked goods, and in soups and stews.	
Farro	This ancient wheat has been cultivated over a millennium. It has a nutty flavour and chewy texture, and is often used in Mediterranean cuisine, such as risotto, salads, and soups or added to baked goods for added texture and flavour. Farro retains its al dente texture long after cooking, used as a base for grain bowls, but requires overnight soaking before cooking.	
Freekeh	Also known as frikeh or farik, is a type of roasted green wheat that is commonly used in Middle Eastern and North African cuisines. It is made by harvesting wheat while it is still green, and then burning off the chaff and straw. The result is a nutty and slightly smoky flavour with a chewy texture. Freekeh can be used in a variety of dishes, such as pilafs, soups, stews and salads, and can be used in both savoury and sweet dishes.	
Khorasan	It is also known as Kamut, an ancient type of wheat possibly originated in the Middle East. It has a rich, buttery flavour and a chewy texture. It is often used in a variety of baked goods, such as bread, pastries, and crackers. It can be added to salads or soups for added texture and flavour.	
Millet	It has a slightly nutty flavour and a fluffy texture when cooked, and comes in different colours: yellow, white, red, or grey. It is used in pilafs, salads, and even baked goods like bread and muffins. In Africa, it is often used to make a type of flatbread called injera, and in India to make roti. Millet is a versatile grain that can be used in both savoury and sweet dishes.	

TABLE 6.2 (Continued)   Major grains and their characteristic features

NAME	REMARKS	IMAGE
Oats	Oats are a highly popular grain with a slightly sweet, nutty flavour and a chewy texture when cooked. In baking, oat flour can be used to make cookies, bread, and other baked goods. Since their bran and germ have not been removed in processing, most food products contain wholegrain oats.	
Quinoa	It is in pseudo-cereal group and is native to the Andes Mountains of South America over 3000 BC. It has nutty-flavoured and used in a variety of dishes: salads, soups, pilafs, and baked goods like bread and muffins, as well in fruits and vegetables.	
Rye	This cereal grain is closely related to wheat and barley and came into cultivation later than wheat, barley and oats. It has a strong, earthy flavour and a dense texture. It is used in many traditional breads, including sourdough, pumpernickel, and rye bread. It is a popular grain in Türkiye, Europe, and is used in Jewish cuisine.	
Sorghum	Sorghum is a type of cereal grain (related to sugar cane and millet) that is able to grow in soils that are poorly nourished, with an unreliable water supply, and primarily grown in Africa and Asia. It has a mild, slightly sweet flavour and a chewy texture. Sorghum flour is used to make chapatis and similar unleavened breads and multi-grain products as well as used in stews.	
Spelt	Commonly grown in southern Germany and Switzerland since 4000 BC. This type of wheat is used until industrialization. It is closely related to wheat and has a nutty flavour and a slightly chewy texture. It is also often used in baking, in breads, pastas, and cakes.	
Teff	This annual cereal grass dates back to 6000 years ago in Ethiopia and Eritrea. It is too small to be milled. It has a mild, nutty flavour and a fine texture, similar to that of poppy seeds. It is used to make injera (a sourdough flatbread) and in baking and is fermented and cooked into a spongy, crêpe-like bread.	
Wheat	It is the most widely cultivated cereal crop and in Turkish dough products. Its varieties are durum wheat (usually to make semolina for pasta and couscous), common wheat (accounts for 80% of worldwide production, is used to make bread), and club wheat (used in pastry). For example, the primary ingredients of Noah's soup (or sweet pudding, asure) are boiled wheat and barley.	
Bulgur	Bulgur (also known as burghul or cracked wheat) is made from durum wheat that is parboiled, dried, and then cracked into different sizes. It is a highly popular ingredient in Turkish, Middle Eastern, Mediterranean cuisines, to make tabbouleh salad, kofte, kibbeh, pilafs, soups, and in stuffing and wrapping. It is easy to cook and has a nutty flavour and a chewy texture.	
Rice	This most popular and staple food comes in many varieties, including white, brown, basmati, jasmine and wild. In white rice, germ and bran are removed, and as in bulgur it is commonly used in Turkish cooking as stuffing ingredients after gently pre-cooked by boiling it in water or stock.	

products or mixed with different whole-grain flours to create the final dough. Bread flour, traditionally made from stone-ground whole wheat, encompasses the bran (outermost layer rich in fibre and B vitamins), the germ (interior comprising oils, vitamins, proteins, minerals, and antioxidants), and the endosperm (interior containing carbohydrates and protein). The endosperm constitutes approximately 85% of the whole wheat grain, and when whole grains are utilized in breadmaking, they should be minimally processed to retain the bran, germ, and endosperm.

To achieve a smoother texture and prolong shelf life, the bran and endosperm are often separated from wheat, resulting in different types of flour used to craft diverse dough products. However, factors such as the availability of specific flour types, evolving eating habits, and even social status have influenced the selection of flour in breadmaking and other dough products. Texture, taste, and cooking time also contribute to the choice of specific flour types, which generally represent refined forms of the same grain.

6.2.1 Flour Types and Quality

Table 6.2 emphasizes the significance of wheat as the primary grain used in producing various flour types for Turkish dough products. The quality of wheat is influenced by factors such as variety, locality, growing conditions (including weather, agricultural practices, and fertilization), harvesting method, and post-harvest treatment and storage. When aiming for an optimal end product like bread, desirable characteristics of wheat flour include better puffing, lower resistance, reduced shrinkage, and appropriate baking time.

The milling process plays a crucial role in flour quality, as it involves removing the bran and reducing the endosperm to small and fine particles. The gradual removal of the outer layers of grains, known as debranning, also affects flour quality and the final outcome, with varying impacts depending on the grain type. If a product claims to be 100% whole wheat flour, all three parts of the grain (bran, germ, and endosperm) should be present in the same proportion as found in the original grain.

While whole wheat flour is more nutritious and preferred for sourdough bread, refined white wheat flour is considered the best for baking bread and is commonly used in some Turkish breads. However, the germ and bran are often added back into refined white flour to obtain various flour types suitable for specific bread products.

For instance, "All-purpose white flour" is produced by milling the endosperms of 80% hard red wheat and 20% soft red wheat. "Bread flour" is milled from hard red spring wheat and utilizes only the endosperms. "Cake flour" is typically derived from soft white wheat and also uses only the endosperms. On the other hand, "pastry flour" requires a higher percentage of protein than cake flour, so it utilizes hard white wheat. It's important to note that "self-rising flour" is a combination of all-purpose flour, baking powder, and salt, although baking powder may not be as effective in humid climates. [1,2] (Figure 6.1).

Moreover, the particle size of wheat flour significantly affects the chemical and rheological properties of bread dough, including deformation and flow characteristics. Smaller particle sizes result in higher levels of wet gluten, sedimentation value (indicating the quantity and quality of gluten), and falling number (a measure of starch chain integrity) [3]. Interestingly, larger particle sizes in flour lead to longer dough development time and increased dough stability.

It is important to recognize that flour quality for milling and baking is subjective and can vary depending on several factors. However, certain characteristics such as protein quantity and quality, starch damage and amylase content are considered critical for technical quality. Mills and bakeries seek consistent and stable wheat and flour quality for reliable outcomes.

Wheat hardness is a significant quality descriptor used to classify wheat grains as soft, hard, or durum. Wheat grains can also be categorized into two major groups based on protein percentage: those with the highest protein content and those with the lowest. Different types of wheat are used for various baked goods, including bread, rolls, cakes, cookies, crackers, pastries, and muffins.

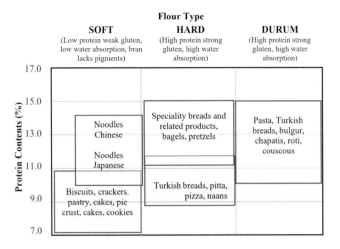

FIGURE 6.1   Flour types with wheat classes/protein contents and end-products.
Ertuğrul, Nesimi, Personal drawing (compiled from [4] and [5–12]).

Ensuring uniformity in bread production necessitates a consistent milling process and a uniform dough-making process. Achieving accurately controlled heat sources can be challenging, particularly in wood-fired ovens. Experience and experimentation with the oven's heat transfer characteristics and the quality of firewood can aid in achieving the desired results.

Ideally, it is preferable to have whole-grain flour with minimal processing. While claims suggest that heat generated from steel mills may harm the nutritional value of flour, this has not been proven. Stone-ground whole wheat flour is a common method, and the wear of grinding stones over time does not impact the quality of the flour or resulting dough.

6.2.2 Dough Making Process

Due to the shared characteristics of dough making for a wide range of bread products, the upcoming chapter will delve into leavening methods and associated practices. Turkish breads and dough products predominantly utilize grains such as wheat, oats, barley, millet, corn, and rye. Additionally, ingredients like potato and chickpea flour, dried pumpkin rinds, and chestnuts can be combined with wheat flour to create a thick, malleable, and occasionally elastic dough paste.

Basic Turkish breads adhere to a simple formula comprising four ingredients: wheat flour, yeast (as a leavening agent), salt, and water. The dough can be leavened (with yeast) or unleavened. Leavened dough results in bread with a rustic sweetness and a natural grain flavour, while unleavened dough remains thin or flaky, yielding a plain taste, as found in certain Turkish breads. Optional ingredients such as sugar, yoghurt, milk, butter, and olive oil can be incorporated into the dough, although it is important to note that additional sugar can prolong fermentation times and contribute sweetness to the final product.

Flatbreads, including pide, employ whole-wheat flour or very high-extraction white flour with high absorption. When baked at high temperatures, steam and gas generated by leavening agents create a void in the centre of the bread. Inadequate gas retention leads to low volume and a dense structure.

The consistency of the dough in Turkish breads varies based on factors such as the blend of flour types, water and ambient temperature, effectiveness of mixing, and resting time. The typical ratios for stiff and soft dough are as follows (see Chapter 7 for a detailed summary of the dough substances):

- Stiff dough: 1 cup water, 4 cups flour mix
- Soft dough: 1 cup water, 3 cups flour mix

Over time, dough-making processes have been developed to achieve flavourful and visually appealing end products that are consistent and reproducible. While the final outcome depends on various factors, two principal dough-making processes and their steps are illustrated in Figure 6.4.

The protein in wheat flour possesses the ability to form a dough that retains gas, a critical aspect for leavened dough-based products. Yeast (in the form of active dry, fresh, liquid, and instant) consumes sugars, producing carbon dioxide and ethanol, which contribute to gas production in the dough. The gas expands air cells, resulting in the cellular crumb structure found in bread products. Salt serves two primary purposes: as a flavouring agent and to aid in gas retention. Unsalted dough tends to be bland and generally lacks palatability.

Dry yeast can activate at water temperatures warmer than 43°C, so it should be kept sealed and dissolved into a liquid with sugar and milk before adding it to a recipe. Fresh yeast should be stored in the fridge and used within a couple of weeks of purchase. Traditional Turkish bread yeasts, such as sourdough bread yeast, often come in the form of a slurry consisting of live yeast organisms, flour (or other carbohydrates), and water.

It is important to note that the straight-dough process shown in Figure 6.2, left, typically yields end products that are less flavourful compared to other processes. This is primarily due to shorter fermentation times and a lower variety of flavour compounds used. Straight-dough processes also require carefully maintained production schedules. In contrast, sponge-dough processing (Figure 6.2, right) allows for some variation in fermentation time, a practice commonly employed in traditional Turkish bread making.

The process flow shown in the figure demonstrates the baking process in a wood-fired oven, which involves uniformly heating dough, whether in the form of sheets or moulded and deposited into baking pans. During baking, a crust forms, the gas-retaining dough transforms into an air-continuous crumb, flavour reactions occur in both the crust and crumb, and uniform browning reactions take place in the crust. To achieve this, it is important to avoid direct flame-radiated heat.

In pan-bread baking, the recommended floor temperature in a wood-fired oven ranges from 190 to 230°C, and it typically takes around 15–25 minutes to bake approximately half a kilogram of dough. However, in Turkish pide (plain or with fillings), the temperature ranges between 270 and 300°C, and the baking/cooking time is shorter, around 10–15 minutes.

The specific bake time and temperature can vary depending on the type of flour used and the duration of dough preparation. In some cases, spraying water on the dough surface or placing a tray with water inside the oven may be necessary to achieve a more viscous starch gel and a thicker final crust. Certain Turkish flatbreads, such as yufka, require higher temperatures to facilitate rapid crust formation and the creation of larger steam pockets. Spraying water on the top layer of the flat-shaped dough creates a temperature difference between the top and bottom layers, resulting in bigger pockets if desired.

Frequent practice and an understanding of the thermal behaviour of the wood-fired oven, often through measurement, are crucial. Generally, lean formulas with low fat and sugar content require higher bake temperatures and shorter bake times compared to rich formulas that incorporate ingredients such as milk, yoghurt, oil, and sugar.

It is important to note that the baking process for Trabzon Bread, a type of Turkish bread, involves slow-cooking at a temperature of around 100°C for up to 24 hours, similar to pumpernickel bread. Hence, optimal baking conditions are highly dependent on the ingredients, preparation process, environment, and the specific product, and may require experimentation on a small scale to determine.

As a general guideline for dough preparation in Turkish breads, the dough should have a softness similar to that of an earlobe. A well-developed dough should be stretchable into a thin sheet that can hold its shape. Further details about the dough preparation will be given at the beginning of Chapter 7.

While commercial proofing for bread is typically conducted in controlled environments with high temperature and humidity, such as around 43°C and 85% relative humidity, household proofing can be achieved by keeping the dough-ball trays at room temperature and covering them with a proofing cloth, provided that the water temperature is warm enough.

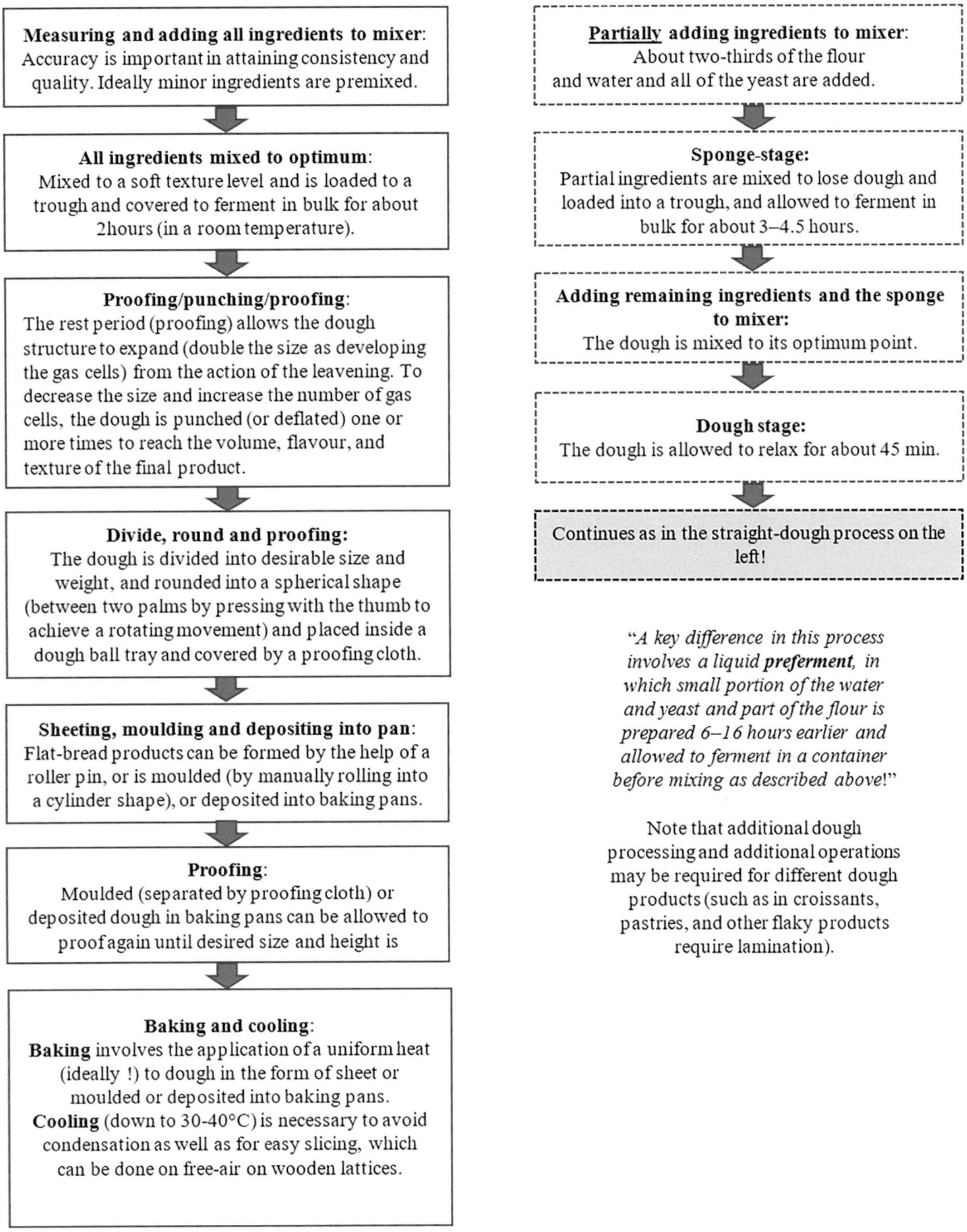

Measuring and adding all ingredients to mixer:
Accuracy is important in attaining consistency and quality. Ideally minor ingredients are premixed.

↓

All ingredients mixed to optimum:
Mixed to a soft texture level and is loaded to a trough and covered to ferment in bulk for about 2 hours (in a room temperature).

↓

Proofing/punching/proofing:
The rest period (proofing) allows the dough structure to expand (double the size as developing the gas cells) from the action of the leavening. To decrease the size and increase the number of gas cells, the dough is punched (or deflated) one or more times to reach the volume, flavour, and texture of the final product.

↓

Divide, round and proofing:
The dough is divided into desirable size and weight, and rounded into a spherical shape (between two palms by pressing with the thumb to achieve a rotating movement) and placed inside a dough ball tray and covered by a proofing cloth.

↓

Sheeting, moulding and depositing into pan:
Flat-bread products can be formed by the help of a roller pin, or is moulded (by manually rolling into a cylinder shape), or deposited into baking pans.

↓

Proofing:
Moulded (separated by proofing cloth) or deposited dough in baking pans can be allowed to proof again until desired size and height is

↓

Baking and cooling:
Baking involves the application of a uniform heat (ideally !) to dough in the form of sheet or moulded or deposited into baking pans.
Cooling (down to 30-40°C) is necessary to avoid condensation as well as for easy slicing, which can be done on free-air on wooden lattices.

Partially adding ingredients to mixer:
About two-thirds of the flour and water and all of the yeast are added.

↓

Sponge-stage:
Partial ingredients are mixed to lose dough and loaded into a trough, and allowed to ferment in bulk for about 3–4.5 hours.

↓

Adding remaining ingredients and the sponge to mixer:
The dough is mixed to its optimum point.

↓

Dough stage:
The dough is allowed to relax for about 45 min.

↓

Continues as in the straight-dough process on the left!

*"A key difference in this process involves a liquid **preferment**, in which small portion of the water and yeast and part of the flour is prepared 6–16 hours earlier and allowed to ferment in a container before mixing as described above!"*

Note that additional dough processing and additional operations may be required for different dough products (such as in croissants, pastries, and other flaky products require lamination).

FIGURE 6.2 The steps of "straight-dough process" (left) and additional steps in a "sponge and dough process" (right).

Ertuğrul, Nesimi, Personal drawing.

6.2.3 Final Bread Appearance/Flavour/Texture and Problems

One of the most desirable characteristics in baked products is a golden-brown crust, which is influenced by various factors such as temperature, moisture, pH level, sugar content, and protein or amino acid content. However, altering the cooking time or temperature can impact browning and may also affect other quality attributes such as texture. Quick browning may result in an undercooked product, while reducing browning can lead to a less visually appealing appearance. The smoothness and shine on the surface of baked goods are related to starch gelatinization, which softens the texture and enhances palatability.

Staling is a phenomenon that affects the texture and flavour of bread products, occurring in both the crumb and crust. Freshly baked bread from a wood-fired oven is nearly sterile but can be easily contaminated. If not consumed quickly, most Turkish breads can be stored in a freezer after packaging, and when reheated, they can regain their hard, crispy crust. However, this process can only be repeated a limited number of times before the bread becomes dehydrated and overbaked.

While browning is a desirable characteristic in certain Turkish bread products, the outer layers are often brushed with egg yolk, egg/yoghurt mix, or a diluted flour mixture (a cost-effective solution practiced in commercial ovens) before baking to achieve a relatively uniform browning appearance. These washes are frequently garnished with white and black sesame seeds or poppy seeds, which enhance the flavour.

To increase the shelf life and modify the texture and taste of bread, some Turkish bread products incorporate potato flour (or mashed potato) or chickpea flour (or cooked chickpeas with the bran removed) into the dough. The typical ratio is 1/4 potato flour combined with 3/4 all-purpose flour (or whole wheat flour). However, chickpea flour does not provide much rise to the dough due to the lack of gluten, so the ratio of chickpea flour needs to be adjusted to achieve the desired texture. For quick breads (unleavened) and yeast breads (leavened), the ratio can be 50%.

While having a fundamental understanding is important, experimenting with a product in a specific wood-fired oven leads to better outcomes. Table 6.3 provides a summary of common problems in bread cooking, including their primary causes and potential mitigation measures that can be considered during formulation and processing.

It is important to note that chemical leavening agents have an impact on both the flavour and texture of soft wheat products and typically contain acids and bases. The reaction rate, and thus the gas evolution rate, can be controlled by influencing the reactivity of the chemical substances, which often include sodium bicarbonate as the primary source of leavening gas carbon dioxide. Potassium bicarbonate is also used to reduce the sodium content. An acidulant is combined with the bicarbonate source, lowering the pH level of the dough mix and causing the release of gaseous carbon dioxide while imparting a tart, sour, or acidic flavour to the food and enhancing the perceived sweetness of the bread.

Although it will be explained in detail in Chapter 7, there are three main types of leavening agents: biological (yeast), chemical, and steam (mechanical). Yeast consists of single-celled organisms (a type of fungus) and is responsible for the fermentation process, in which the biological leavening agent consumes sugar and produces carbon dioxide (CO_2) and alcohol.

Active dry yeast requires activation, or "proofing," by dissolving it in warm water (around 43°C) before mixing it into the dough. In contrast, instant dry yeast can be added directly to the dough without proofing and requires a smaller amount compared to active dry yeast. Fresh yeast can also be used, either homemade or store-bought, and can be added directly to the dough or dissolved in water (see further details in Chapter 7).

It is important to note that adding more yeast to the dough mix will not cause the bread to rise more, but it will result in a more pronounced yeast flavour.

Chemical leavening agents are commonly employed in baked goods to achieve the desired texture and flavour. Baking soda, also known as sodium bicarbonate, is a commonly used chemical leavening agent with a pH level of 8 to 9 (indicating it is a base). When combined with an acidic ingredient such as buttermilk, lemon juice, yoghurt, sour cream, molasses (grape, carob), or honey, baking soda reacts immediately and releases carbon dioxide. Baking powder, on the other hand, is a mixture of baking soda and a powdered acidic ingredient. It reacts when moistened and further reacts when heated during the baking process.

TABLE 6.3 Some common problems in appearance, texture, flavour and processing, causes, and mitigation measures (adapted from [5])

Particular Appearance	Causes	Mitigation Measures
Too brown	High sugar contents	Reduce sugar level. Check flour α-amylase and/or starch damage. Avoid flame-radiation/reduce temperature.
	High heat in short time	Reduce bake time.
	Long bake time	Increase water amount.
	Low moisture	Spray water before baking.
	High pH	Check dough pH level.
Blistered surface	Air pockets under dough surface	Adjust molder. Dock dough. Reduce dusting flour.
Voids in crumb	Air pockets in dough	Check rolling, moulding, or folding process operations. Reduce dusting flour
Dull surface	Underdeveloped starch gel on surface	Include a boiling water in a tray Use moist fire wood (in Turkish pide)
Particular Texture	**Causes**	**Mitigation Measures**
Low volume or dense crumb	Poor gas-holding properties	Adjust mix time. Check post-mixing work Adjust flour and/or water content. Change flour type. Eliminate reducing agents. Add oxidant, dough strengtheners and/or wheat gluten.
	Inadequate leavening	Increase yeast, sugar and/or chemical leavening content. Check correct balance of acidulant and bicarbonate.
	Structure does not set in expanded form	Remove gluten. Use weaker flour. Add malt and/or reducing agents. Check oven temperature. Increase bake time.
Firm crumb	Starch interactions	Add shortening. Add mono- and diglycerides
Elastic crumb	Gluten interactions	Reduce vital wheat gluten. Use weaker flour. Add shortening.
Particular Flavour	**Causes**	**Mitigation Measures**
Bland	No salt or low salt	Adjust salt level
Sour	Lactic acid bacteria	Check yeast for contamination.
Cheesy	Protease activity	Reduce fermentation time.
Fruity	Yeast type	Change yeast type. Reduce fermentation time
Bitter	Browning reactions	As listed above

(Continued)

TABLE 6.3 *(Continued)* Some common problems in appearance, texture, flavour and processing, causes, and mitigation measures (adapted from [5])

Processing Issues	Causes	Mitigation Measures
Malformed product or tearing	Viscoelastic properties	Adjust mixing time, allow more rest time. Reduce directionality in moulding and sheeting operation. Adjust flour strength. Add dough strengthened or reducing agents

Steam is another leavening agent used in certain baked goods. When the water in the dough reaches temperatures of 100°C and above, water vapour is produced, resulting in a significant increase in volume. Examples of baked goods that utilize steam as a leavening agent include steam buns and puff pastry.

6.2.4 Bread Types

Anatolia, which encompasses present-day Türkiye, holds a significant place in the history of bread, boasting a remarkable diversity of bread types. Table 6.4 provides an overview of some commonly found Turkish breads in the region, highlighting their distinct characteristics, shapes, and appearances. These breads are typically known by concise, one-word names. In Anatolia, neighbouring provinces often have variations of the same bread, distinguished by local names.

While the breads listed share similarities, they can also be categorized based on whether they are leavened or unleavened, their shapes (circular, oval, oblong, triangular, or rectangular), and their styles (flat, long/large rounded, half-spherical, or ring-shaped). In addition, certain Turkish breads are plain, while others are filled, such as pide and lahmacun. As mentioned previously, many Turkish breads feature an egg wash or yoghurt wash, and topped with black or white sesame or poppy seeds.

Further sub-classifications can be made based on flour types, baking platforms (wood-fired oven, tandır, or hearth-cooked on a sac), cooking surfaces (directly on the oven floor or wall), or cooking utensils (using a metal cooking tray or clay pot with or without a lid). These variations contribute to the rich and diverse landscape of Turkish bread culture.

In addition to the breads and pastries listed in Table 6.4, there is a wide variety of other dough products and breads commonly consumed in Türkiye, each with its own unique local name. Some notable examples include nohut ekmeği (chickpea bread), peksimet, bileki (pileki), çalı ekmeği (bush bread), esmer ekmek (brown bread), galeta, gevrek, haşhaşlı ekmek (poppy seed bread), lokum ekmeği (delight bread), tandır ekmegi (hearth bread, usually named after the local province), açma, bezirme, biliğ, bohça, boyos, cadi, çan bezesi, çarşı ekmeği (market bread), çavdar ekmeği (rye bread), çerkez ekmeği, cızlama, çukmin, dızmana, eksili ekmek (sour bread), fetil, fıtcın, francala, gartalaş, gevrek, girde, gömme gözleme, güdül, habiycın, halka (round bread), hamursuz, işkefe, kabartlama, kalın, katlama, katmer, kayasa, kelecoş, kete, kirde, köy ekmeği (village bread), küncü, kuru ekmek, nokul, otlu ekmek (herb bread), övme, papara, peremeç, pezdirme, bezdirme, pırasalı ekmek (leek bread), pişi, pıt pıt, pobuç, poğaça, sac ekmeği, sacaltı, şebit, gardalaç, seyme, şıllık, sinçü, soğanlı ekmek (onion bread), tablama, tandır gevregi, tepsi ekmeği (tray bread), ter ekmeği (sweat bread), and toraman.

Two widely enjoyed flatbread dishes in Turkish cuisine are pide and lahmacun, which serve as the bases for flavorful and diverse dishes with rich fillings.

Pide is typically oval-shaped (see Figure 6.3) with dimensions of length ranging from 30 to 80 cm, width of 10 to 20 cm, and a thickness between 5 mm and 15 mm. The crust of pide is relatively soft and

TABLE 6.4 Turkish breads practiced in Anatolia and their characteristic features

NAME	LEAVENED OR UNLEAVENED	CHARACTERISTIC FEATURES	TYPICAL SHAPES AND APPEARANCES
Bazlama (Also known as Bazlamaç, Bezdirme, Bezirme)	Leavened Or Unleavened	This flatbread is made with wheat flour, yeast, salt, and water. It is similar to pita bread or naan, but without a pocket. The dough is rolled out into circles (typically in 10–15cm diameter) or ovals and with a common thickness less than 1cm. It is typically eaten with a variety of dishes, such as meat, cheese, or vegetable-based stews, and is often served as a breakfast item, alongside tea and cheese in Türkiye.	
Millet Bazlama	Yeast	It is a type of bread made from millet flour, a variation of the traditional Turkish bazlama bread. It is traditionally cooked on a sac, which is a convex metal griddle, and is typically served as a side dish or used as a base for toppings such as cheese, vegetables, or meat.	
Çerepene	Yeast	It a type of flat bread that is traditional to the Black Sea region of Türkiye. It is made with a mixture of cornmeal and wheat flour, and is typically cooked on a griddle or in a wood-fired oven, but inside a clay pot known as "çerepene" or in a metal round tray either lid open. It is usually served as an accompaniment to local dishes such as stews and soups, and is also eaten with cheese, honey, or butter. Its texture is slightly crispy on the outside and soft and chewy on the inside.	
Ebeleme	Yeast	Particularly popular in central Anatolia and in the province of Gaziantep/Türkiye. It is a round (bazlama size), flatbread (2–5 mm thick) that is made with a mixture of wheat flour and cornmeal, and is typically baked in a wood-fired oven or on sac. It is cooked over high heat until it is golden brown and slightly crispy on the outside, while remaining soft and fluffy on the inside. After cooking both side can be brushed with butter. It is often served with kebabs, stews, and salads or with cheese topping.	

(Continued)

TABLE 6.4 (Continued)   Turkish breads practiced in Anatolia and their characteristic features

NAME	LEAVENED OR UNLEAVENED	CHARACTERISTIC FEATURES	TYPICAL SHAPES AND APPEARANCES
Fodula or Fodla	Leavened	This bread is made from wholemeal flour, thinner than pita and in a rectangular or round shape, with crispy outer layer and softer inside, about 300 gr weight. It was a common bread type for the Janissaries (an elite infantry division of the Ottoman Empire) and other palace workers. Recently, its thicker and round versions are served after stuffing.	
Gilik	Leavened	It is a round bread with diameters of about 7cm and 20 cm, prepared with a wide hole and flat in the middle. from leavened dough, and sprinkled with sesame seeds. Gilik is usually prepared for two purposes: for a religious day (bagels up to 7 cm in diameter) and for the memory of a loved one to distributed to the neighbours, in Sivas and Erzincan provinces of Türkiye.	
Gömeç (Also known as Cörek, gömbe, göbe and göbü)	Leavened or Unleavened	It is a type of bread that is traditionally made in the town of Gömeç in the Balıkesir province of Türkiye. It is made using a sourdough starter and is baked in a wood-fired oven, which gives it a crispy crust and a soft and chewy interior. A mix of cumin hay and sawdust used as a heat source in the oven, known as "saçkı," and cooked for about 10 min on a metal tray while the oven door is closed. It has a unique flavour and is typically shaped into a round loaf and is scored with decorative pattern before baking. Served with meals or used as a base for sandwiches.	
Gübaye	Leavened	It is made from a mixture of wheat flour and cornmeal. The bread is leavened with a sourdough starter and is baked in a wood-fired oven, which gives it a crispy crust and a slightly smoky flavour. If cooked longer it can last longer, then has to be moistened before eating. It is a traditional bread in the Adana province of Türkiye. It is often served with meals or used as a base for sandwiches.	

TABLE 6.4 *(Continued)* Turkish breads practiced in Anatolia and their characteristic features

NAME	LEAVENED OR UNLEAVENED	CHARACTERISTIC FEATURES	TYPICAL SHAPES AND APPEARANCES
Halka	Leavened	It is made from a mixture of wheat flour, water, yeast, and salt. The dough is made into a ring shape (as a bagel or a donut but much larger) and is boiled briefly before being baked. It has a unique texture and flavour, making it slightly chewy on the inside and crispy on the outside often served with meals or used as a base for sandwiches, and in breakfast paired with cheese, olives, and tea. If cooked longer and allowed to dry it can last longer. It is a traditional bread in the city of Gaziantep in Türkiye.	
Kakala	Leavened	It is a flatbread made from a mixture of cornmeal, wheat flour, and water. It is cooked in a deep stone cup called bileki. The result is a crispy, slightly charred exterior with a soft and chewy interior. This bread is often served with a variety of dishes, including stews, soups, and grilled meats. It is a traditional bread originating from Artvin and Black Sea region of Türkiye.	
Lavaş (also known as lavaj and lavas)	Unleavened	It is made from a mixture of flour, water, and salt, which is kneaded into a dough and rolled out thinly before being cooked on a sac or in a wood-fired oven. It is usually prepared in a round shape (about 30cm in diameter) flat-bread (about 1-cm thick). It has a soft and slightly chewy texture, with a delicate flavour. It can pair well with a variety of foods, as a wrap for kebabs and falafel, as a side with dips and spreads.	
Mayalı Şepe	Leavened	The bread is made with a combination of wheat flour and rye flour, and it is known for its dense texture and hearty flavour. It is a sourdough bread using a natural starter culture, which gives a distinctive sour flavour. Usually prepared in a circle shape (between 10–40 cm diameter). When baked in a wood-fired oven, it gives a crispy crust and a smoky flavour. It is specific to the Konya province in Türkiye.	

(Continued)

TABLE 6.4 (*Continued*)   Turkish breads practiced in Anatolia and their characteristic features

NAME	LEAVENED OR UNLEAVENED	CHARACTERISTIC FEATURES	TYPICAL SHAPES AND APPEARANCES
		Recently, its ingredients include various fresh herbs and edible plants such as parsley, spring onion or sideritis plant (madimak). Note that sideritis is native to Türkiye, Greece, and Albania and has a strong, earthy flavor and is often used to add a unique taste to soups, stews as well as teas.	
Somun (Loaf)	Leavened	It is typically made with flour, water, salt, and yeast. It is one of the most popular types of bread in Türkiye as well as other neighboring countries in the region, with a long round shape. It is known for its crispy crust and soft interior, and it is often used to make sandwiches or served as a side dish to a variety of meals.	
Pebble-Baked Bread	Leavened	This flatbread is typically made with whole wheat flour, water, yeast, and salt. The dough is allowed to rise before being shaped into long, oval-shaped flat bread with about 1 cm thick, which are then placed onto a bed of small, smooth pebbles that are preheated in the wood-fired oven. When baked on the pebbles, a unique texture is created and with flavor in the crust. It is known as known as Sangak bread in Iran and also popular in Afghanistan and Azerbaijan.	
Tava Ekmeği (Pan Bread)	Leavened	It is typically made with a combination of wheat flour, yeast, water, salt, and sometimes with olive oil. In some regions, chickpea flour is also used. The dough is rolled out thinly and then baked in an oiled pan (or metal tray) until it becomes slightly crispy on the outside and soft on the inside. It is a popular bread in Türkiye and is often served as an accompaniment to a variety of dishes, such as grilled meats, stews, and salads.	
Yuvarlak (Round Bread)	Leavened	It is a simple bread made with a combination of flour, water, yeast, and salt. Dough is shaped into a round shape (10–20 cm diameter). When baked it becomes golden	

TABLE 6.4 (Continued) Turkish breads practiced in Anatolia and their characteristic features

NAME	LEAVENED OR UNLEAVENED	CHARACTERISTIC FEATURES	TYPICAL SHAPES AND APPEARANCES
		brown and crispy on the outside and soft on the inside. It is served with various dishes such as kebabs, stews, and salads.	
Dari Ekmeği (Corn Bread)	Unleavened	It is a type of bread made using cornmeal or flour made from the grain sorghum, which is also known as dari in some regions. The bread made from sorghum flour mixed with wheat flour becomes denser and less elastic texture compared to pure wheat bread. However, if bread made solely from sorghum flour becomes less fluffy and has a shorter shelf life.	
Yufka Ekmeği (Yufka Bread)	Unleavened	It is made from flour, water, and salt, and rolled out into thin circles (similar to phyllo dough) which are then cooked on a griddle or in a wood-fired oven. The result is a thin and crispy bread that is commonly used in Turkish cuisine for dishes such as dürüm, a type of wrap filled with meat and vegetables.	
Yufka for börek	Unleavened	The primary difference from the previous mixture is that dough is usually formed by an egg. Yufka dough is similar to phyllo dough, and it can be rolled out by hand or using a special rolling pin called an "oklava." Yufka is usually formed by eggs and it is thicker and more pliable, making it easier to work with when making savoury dishes named börek. A range of börek can be prepared using such yufka, such as "su böregi" (using boiled yufka before baking or frying), and börek with cheese, spinach, ground meat or potatoes. Börek is often served with tea or as part of a mezze platter.	
Filo Yufka (Phyllo Pastry)	Unleavened	Filo yufka is made from a mixture of flour, water, and a small amount of oil. The dough is kneaded until it is smooth and elastic, and then rolled out into very thin, delicate sheets. These sheets are typically used as a pastry dough for making dishes like	

TABLE 6.4 (Continued)   Turkish breads practiced in Anatolia and their characteristic features

NAME	LEAVENED OR UNLEAVENED	CHARACTERISTIC FEATURES	TYPICAL SHAPES AND APPEARANCES
		baklava, borek, and other savoury or sweet pastries. It is known for its delicate, flaky texture and its ability to create many layers when stacked together. It is often brushed with melted butter or oil between the layers to help create a crispy, golden-brown crust when baked. It can be used as a pastry dough for making pies and pastries, as a wrapper for savoury fillings, or as a crispy topping for various dishes.	
Pide Base	Leavened	Pide is the most popular Turkish bread, and its dough is made from a mixture of flour, water, yeast, and salt. It is usually shaped into a round or a long, oval shape with a slightly raised edge. The pide can be plain or topped with a variety of ingredients, such as cheese, meat, vegetables, and herbs, and then baked in a wood-fired oven until the crust is crispy and the toppings are cooked through.	
Lahmacun Base	Unleavened or leavened	The dough for lahmacun is made from a mixture of flour, water, and salt, and it is rolled out very thinly before being topped with a thin layer of mixture of ingredients prior to cooking. Its dough is made unleavened if wanted crusty base and leavened if soft base is preferred. Lahmacun is typically served with a variety of toppings, which will be covered in the next subsection. Although it can be the main dish together with the rich variety of toppings, it can also be served as starter of appetizer.	
Simit	Leavened	It is a type of circular bread that is a popular snack in Türkiye. Simit is made from a dough that is usually made from flour, water, yeast, and a small amount of sugar and salt. The dough is formed into a long rope shape, twisted into a circular	

TABLE 6.4 *(Continued)*   Turkish breads practiced in Anatolia and their characteristic features

NAME	LEAVENED OR UNLEAVENED	CHARACTERISTIC FEATURES	TYPICAL SHAPES AND APPEARANCES
Kömbe (Also known as Kömeç, Kömme and Gömbe)	Unleavened Leavened	shape, and then dipped in a mixture of water and grape molasses before being coated in sesame seeds and baked. The grape molasses gives simit its distinctive dark colour and slightly sweet flavour, while the sesame seeds add a nutty and crunchy texture. It is usually eaten as a snack, either on its own or with tea or coffee, Kömbe bread (ash-pitta) is made in many parts of Anatolia from unleavened and leavened dough that contains wheat flour, water, and with/without salt. Multiple layers of yufka style dough is spread on a tray with available ingredients such as spinach, minced beef, kavurma*. It is baked by burying in the sparking ash fire of the newly extinguished oven or hearth.	

Notes
* **Kavurma**: It is a Turkish dish made from cubed-meat (lamb or beef) that has been slow-cooked in its own fat until it becomes tender and flavorful. It is used in stews, soups, rice dishes, cold side dish, and as fillings.

fluffy, and it is sliced into bite-sized pieces before serving. In the Black Sea region of Türkiye, pide may be round in shape, but the oval shape is more common. There is also a closed version of pide (see Figure 6.3) where the sides of the dough base are joined together to enclose the filling completely before baking.

Lahmacun, on the other hand, is a thin and crispy flatbread (3–5 mm) topped with a mixture of ground meat (usually lamb or beef) and various other toppings such as tomatoes, onions, parsley, mint, garlic, and red pepper flakes. It has a round and uniformly flat shape (see Figure 6.3), with a diameter typically ranging from 12 to 20 cm. The smallest version of lahmacun, with a diameter of approximately 8 cm, is known as "findik lahmacun" or "tiny/small lahmacun." Lahmacun is often served as a snack or appetizer and is enjoyed with a squeeze of lemon juice for added flavour, or garnished with a range of vegetables as illustrated in Figure 6.4.

Table 6.5 provides an overview of the ingredients and garnishes used in pide and lahmacun, as well as the common types of pides and lahmacuns in Türkiye. The various combinations of ingredients listed in the table form the toppings of specific pide types, and in many cases, a local town/city name is given to a pide that contains identical ingredients.

6.3 WRAPPING AND STUFFING: LEAVES, VEGETABLES/ FRUITS AND OTHERS

"Dolma" and "sarma" are popular dishes frequently included in Turkish cuisine, and they are also commonly enjoyed in other countries throughout the Middle East and Eastern Europe. The word

(a)   (b)   (c)

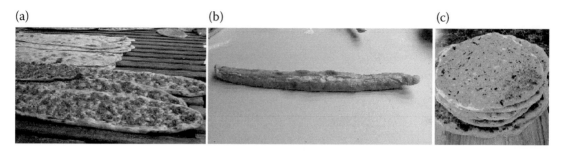

FIGURE 6.3   Typical shapes of open-oval shape (a) and closed (b) forms of pides, and typical lahmacun (c).

Ertuğrul, Nesimi, Personal photograph, Adelaide, Australia, 2022.

FIGURE 6.4   Some of the garnishes used in lahmacun, whole and chopped: parsley, yellow/green/red capsicum, lemon, carrot, turnip, Jerusalem artichoke, tomato, cucumber and spring onion.

Ertuğrul, Nesimi, Personal photograph, Adelaide, Australia, 2023.

"dolma" in Turkish means "stuffed," and the dish typically involves filling vegetables or fruits with a flavourful mixture of ingredients. In contrast, "sarma" means "wrapped," and it entails wrapping edible leaves around a filling that can include a variety of ingredients, similar to "dolma." Both dishes are usually cooked in a wood-fired oven and are often served as either appetizers or main courses.

The specific techniques for wrapping and stuffing used in "dolma" and "sarma" can vary depending on the region and the recipe. However, there are significant similarities in the ingredients used for both types of dishes, with minor variations reflecting local and traditional differences. Additionally, the spices and herbs used in the stuffing and wrapping can vary based on regional preferences or the food culture of a particular area.

Table 6.6 provides an overview of the possible ingredients used for wrapping and stuffing in Turkish cuisine, which can be customized to suit a wide range of tastes and dietary preferences. By combining these ingredients, various mixtures for raw or lightly pre-cooked stuffing and wrapping can be created. Figure 6.5 illustrates the texture and appearance of some selected ingredients commonly used in stuffing and wrapping. Overall, both "dolma" and "sarma" are flavourful and nutritious dishes that showcase the rich culinary heritage of Türkiye.

It is worth noting that in Turkish cooking, thinly slicing plants and meaty products (known as "emince") is often preferred over crushing or pressing, as it results in a more settled flavour. These cuts can be made lengthwise, crossed, or in a combination to achieve the desired size, texture, and appearance of the ingredients. Note that the same style of medium or fine chopping is also preferred when forming ingredients in pide and lahmacun.

TABLE 6.5   Popular ingredients and garnishes used on pide and lahmacun bases

PIDE INGREDIENTS AND POPULAR TYPES	LAHMACUN INGREDIENTS AND TYPES AND GARNISHES

PIDE INGREDIENTS AND POPULAR TYPES

Ingredients:
Minced or finely chopped beef, veal, lamb, chicken and ox/lamb tongue; feta, halloumi, blue and cheddar cheese; tulum cheese (goat's milk cheese ripened in a goatskin casing); baby spinach; sliced tomato, sun-dried tomato, onion, garlic and mushroom; egg; anchovies; thinly chopped green and black olives; green/yellow long pepper; red/green/yellow capsicum; chopped or sliced, sausage, sucuk, salami, pastirami; fresh basil, oregano, parsley leaves; tomato paste; tahini; olive oil; salt; spices.

Turkish Names (with basic ingredients):
Kıymalı Pide (minced lamb, chopped tomato and onion, black, salt and pepper/spices).
Kuşbaşı Pide (kuşbaşı means diced, cubed, sliced lamb or veal)
Kavurmalı Pide (roasted diced meat).
Dönerli Pide (with döner/yiros).
Develi Cıvıklısı (finely chopped lamb meat from shoulder and ribs, chopped tomato and pepper).
Samsun/Bafra/Karadeniz Pidesi (Closed-type, pre-cooked minced veal or lamb, with/without chopped onion, tomato, mushroom).
Mevlana/Konya Pidesi (minced lamb/veal, blue and cheddar cheese, …).
Bıçak Arası Pide (diced lamb/veal, tomato, pepper, olive oil, …).
Tavuklu Pide (chopped chicken, mushroom, capsicum and tomato, topped with mozzarella and/or feta cheese).
Sucuklu/Yumurtali Pide (sliced sucuk, minced chedar cheese, egg yolk).
Pastırmalı Pide (Turkish pastrami/cured beef, cheddar or feta cheese and egg)
Beyaz Peynirli Pide (Fetta cheese and chopped parsley, egg white, salt, spice)
Ödemiş/ Tongül Pide (Precooked minced meat, black pepper, salt, finely chopped parsley, egg, tulum cheese and melted butter)
Patatesli Pide (cooked potato, onion, tomato paste, black pepper, spices, salt).
Ispanaklı Pide (spinach, fetta cheese)
Karisik Pide (Combination pide, includes choosen ingredients. For example, tomato with cheddar, roasted diced meat with cheddar, roasted diced meat with egg, minced meat/egg).
Tahini Pide (sweet version of pide, cooked with tahini spread over the dough base, and served plain or with cream or crushed walnut and peanuts).
Ballı Tavas Pide (sweet version of pide, tahini is mixed to the dough and honey and walnuts as toppings).
Aksaray Şerbetli Pide (sweet version of pide, prepared with minced meat and cheese but served with a syrup made of lemon, water and sugar).

LAHMACUN INGREDIENTS AND TYPES AND GARNISHES

Ingredients:
The range of toppings in Lahmacun hence ingredients are limited. Typical mixtures as toppings include:
- Minced/ground beef, lamb and veal traditionally with higher fat content; finely chopped tomato, parsley and garlic, salt, chilli.
- As above, but tomato is replaced by finely chopped onion.
- and two vegetarian options
- Mashed potato and finely chopped parsley, salt and spice.
- Cooked lentil and finely chopped parsley, salt and spice.

Garnishes (ideally all freshly prepared):
Following range of garnishes can be used to serve with lahmacun:

Lemon or citrus wedges ready to squeeze, parsley leaves, baby spinach, rocket (Arugula, Italian cress) leaves, thinly sliced red onion usually mixed with sumac, sorrel (Kuzukulağı), stinging nettle (Isırgan), choices of pickles, purslane/watercress (Semizotu), sliced radish, garlic leaves, garlic chives, long/thin sliced carrots and cucumbers, cos lettuce, frisee lettuce, leaf lettuce, longitudinaly chopped spring onion, sliced tomato, and isot.

TABLE 6.6　Ingredients that can used for wrapping and stuffing in Turkish cuisine

- Rice, bulgur
- Crashed nuts (walnuts, hazelnuts, peanuts, pistachio), pine nuts, sunflower seeds, pumpkin seeds
- Finely chopped onion, tomato, celery, mushroom
- Finely chopped parsley, dill, tomato, garlic, basil, mint, or dry versions
- Chestnut, prunes, dates, dry/fresh apricots, small cut apples, quinces
- Sultanas, currants, raisins
- Meat and seafood ingredients (such as beef, lamb, veal, prawn, mussels, sausage, sucuk, oysters, tongue, heart, and giblet) usually in the forms of ground/minced or dice/cube/stew cut or emince (lengthwise and crossed)
- Spices: cumin, black pepper, paprika, cinnamon, coriander, sumac
- Olive oil, butter, tomato paste, salt, caster sugar, sweet/mild/hot red pepper paste
- Pomegranate, carob, grape syrup, and molasses
- Pomegranate seeds
- Carob powder
- Lemon and lemon juice
- Riccotta cheese (lor peyniri)
- Breadcrumbs

To prepare the wrapping, the leaves are first blanched in boiling water to soften them. Then, a combination of ingredients is filled into the leaves, and they are tightly rolled into small cylinder shapes. The rolls are arranged tightly in a pot with a lid and covered with water, fresh tomato sauce, lemon slices, or juice. The pot is simmered slowly until the wrapping is fully cooked and the flavours have melded together. It's important to note that the quality of the dish is defined by the texture of the leaves (whether they are hard or soft, fresh or old) and the ingredients used. Therefore, the fillings used in the wrapping are commonly required to be pre-cooked, which also enhances the final flavour. Most "sarma" dishes are usually served cold or at room temperature, and they can be enjoyed as an appetizer or a main dish.

A highly popular dish in Turkish cuisine involves wrapping wine leaves around a mixture of pre-cooked rice, minced meat (lamb or beef), finely chopped onions, herbs, and spices. Sometimes, the meat is omitted, and a purely vegetarian filling is used. These wine leaves filled with the mixture are then rolled and cooked in a pot until the rice is tender. Due to its popularity, pickled wine leaves are commonly used for wrapping throughout the year, and they have a slightly sour taste compared to fresh green leaves.

Leaves are commonly used in Turkish cuisine as a wrapping for a rich mixture of ingredients. They not only prevent dehydration but also infuse their natural flavour into the dish while adding a fibrous texture. Figure 6.6 illustrates the flavourful young plant/tree leaves that are ideal for wood-fired oven cooking. These leaves serve two main purposes: wrapping various ingredients in a rolled shape to contain them during cooking and consuming as a whole, or using them as an outer layer for steaming and keeping the dish moist.

Although the majority of the leaves shown in the figure are commonly used in Turkish cuisine, banana, lotus, hiso, bamboo, shinno, and carob leaves are included in the figure as they are well-known in Asian and Japanese cooking which suit to the wood-fired oven cooking. Some fusion cooking recipes in the final chapter will accommodate these leaves. It's important to note that most of the leaves shown in the figure are edible, except for banana and bamboo leaves.

Additionally, popular fruits and vegetables suitable for stuffing (dolma) are summarized in Figure 6.7. The key takeaway is that similar ingredients and mixtures are used for both wrapping leaves and stuffing fruits/vegetables. While a few recipes will be provided in the final chapter, the style of the pre-cooked mixture remains somewhat similar for wrapping and stuffing. For example, a typical wrapping and stuffing mixture can consist of rice (1 unit), bulgur (cracked wheat, 1 unit), raisins or currants (1/2 unit), pine nuts, finely chopped parsley, dill, and mint, finely chopped onion, garlic, tomato, salt, and olive oil.

FIGURE 6.5 Some of the ingredients commonly used in stuffings and wrapping: minced meat, finely diced meat, cubed meat, finely chopped onion, bulgur (cracked wheat), currants, dry apricot, sultanas, dry grapes, sun-dried and fresh apricots, pine nuts, peanuts, pumpkin seeds, hazelnuts, pistachio, almond, sunflower seeds, prunes, pomegranate, ricotta cheese (lor peyniri), quince, chesnuts/dates/dry figs, walnuts, three grades of carob powder.

Ertuğrul, Nesimi, Personal photograph, Adelaide, Australia, 2022.

Figure 6.7 showcases the numerous variations of "dolma" in Turkish cuisine, highlighting the country's affinity for flavourful and aromatic ingredients, including spices and herbs. It is worth noting that the preparation of dolma is generally simpler compared to sarma.

To prepare dolma, vegetables or fruits need to be cleaned and made ready for stuffing. The tops of the vegetables or fruits should be cut around the stem to create a lid, and the flesh, seeds, and/or membranes from the centre should be removed. A mixture of ingredients should then be prepared, and if rice or bulgur is included, they should be pre-cooked or soaked in water for about 30 minutes before being mixed with the other ingredients.

Once the filling is prepared, it should be placed inside the vegetables or fruits, and their tops should be closed with their own lids. The stuffed vegetables or fruits should be tightly packed together in a pot and covered with water or broth. To prevent them from floating, a heavy plate or weight can be placed on top. The pot is then ready to be cooked on medium heat in a wood-fired oven.

FIGURE 6.6 The types of plant/tree leaves that can be used for wrapping (known as "sarma" in Turkish) various ingredients: from top left: wine, mulberry, cherry, banana, fig, lotus, oak, labada (efelek/evelik in Turkish), hiso, sassafras, nasturtium, bean, hoja santa, bamboo, shinno, carob, pumpkin flower, green or purple cabbage leaves, chard (pazi in Turkish), quince, leek, linden (ihlamur), spinach (ispanak), tailwort (hodan out), and Japanese morning glory (sabah sefası).

Ertuğrul, Nesimi, Personal photograph, Adelaide, Australia, 2022.

Thanks to the wide variety of ingredients listed in Table 6.6 and the natural flavours of the leaves, fruits, and vegetables, exceptionally flavourful dishes (see Table 6.7) can be created, especially when cooked in wood-fired ovens. Dolma is commonly served as a meze or appetizer in Turkish cuisine but can also be enjoyed as a main course. It is typically eaten at room temperature or slightly warm and

FIGURE 6.7 Images of popular fruits and vegetables suitable for stuffing ("dolma") dishes in the wood-fired oven: from top left to right: green small pepper, capsicum, aubergine, tomato, melon, watermelon, gold nugget pumpkin, buttercup pumpkin, butternut pumpkin, green coconut, coconut shell with/without endosperm, potato, quince, pineapple, apple, pear, onion, mushroom, artichoke (enginar), zucchini, dried small pepper (red or green, sweet or hot), dried aubergine, dried cucumber (in Maras region), and dried zucchini.

Ertuğrul, Nesimi, Personal photograph, Adelaide, Australia, 2022.

traditionally accompanied by a dollop of yoghurt mixed with crushed garlic, a drizzle of olive oil, or a spoonful of tomato and garlic sauce.

6.3.1 Other Dishes

Since the term "stuffed" refers to a cooking technique in which a hollowed-out food item is filled with a variety of ingredients, this method is also utilized in other food products. This technique is commonly practiced and highly regarded in Turkish cuisine, particularly when preparing poultry (such as chicken, fowl, turkey, duck, and goose), game (including pheasant, quail, partridge, wild duck, deer, and rabbit), baby lamb/goat, and fish/shellfish after their cavities have been cleaned and gutted.

TABLE 6.7   Summary of well-known and popular "Stuffed" dishes in Türkiye

STUFFED DISHES

Aleppo-Style Dry Aubergine (Halep Dolması)
Bean-Stuffed Zucchini (Fasulyeli Kabak Dolması)
Celery Root (Kereviz Dolması)
Dried Eggplant (Kuru patlıcan Dolması)
Dried Zucchini (Kuru Kabak Dolması)
Eggplant with Olive Oil (Zeytinyağlı Patlıcan Dolması)
Ermenek-Style Eggplant (Ermenek Dolması
Freekeh-Pickled Cucumbers (Firikli Acur Dolması)
Green Tomatoes (Yeşil Domates Dolması)
Meat-Stuffed Zucchini (Etli Kabak Dolması)
Pistachio-Stuffed Zucchini (Fıstıklı Kabak Dolması)

Stuffed:

Apples (Elma Dolması)
Artichokes (Enginar Dolması)
Aubergine, Aleppo Style (Mülebbes Dolma)
Beef Tripe (İşkembe Dolması)
Carrots (Havuç Dolması)
Carrots (Havuç Dolması)
Celery Root (Kereviz Dolması)
Chicken (Tavuk Dolması)
Dried Aubergine (Kuru Patlıcan Dolması)
Dried Zucchini (Kuru Kabak Dolması)
Aubergine with Meat/with Olive Oil
   (Etli/Zeytinyağlı Patlıcan Dolması)
Fish (Balık Dolması)
Green Peppers (Yeşil Biber Dolması)
Green Tomatoes (Yeşil Domates Dolması)

Stuffed:

Lamb (Kuzu Dolması)Lamb
Intestines (Kuzu Bağırsak Dolması)
Leeks (Pırasa Dolması)
Mackerel (Uskumru Dolması)
Melon (Kavun Dolması)
Mussels (Midye Dolması)
Okra (Bamya Dolması)
Onions (Soğan Dolması)
Potatoes (Patates Dolması)
Pumpkin (Balkabağı Dolması)
Quinces (Ayva Dolması)
Ribs (Kaburga Dolması)
Squash-Cucuzza (Haylan or Asma Kabağı
   Dolması)
Tomatoes (Domates Dolması)
Turnips (Şalgam Dolması)
Zucchini with Meat/with Unsalted Cheese (Etli/
   Tuzsuz Peynirli Kabak Dolması)
Tahini Fruits/Vegetables (Tahinli Dolma)
White Zucchini with Unsalted Cheese (Sakız
   Kabağı Dolması)
White Zucchini with Curd (Katıklı Dolma)

Note: Firik bulgur is known as Yarma in Anatolia, which is considered the ancestor of bulgur. It is obtained from the stage before the ripening and drying of wheat. It has a sooty taste due to boiling and drying process in its preparation.

It is important to note that, although seafood is not generally categorized as "meat" in the same way as land animals, this book includes all edible aquatic animals in the "meat" category. Additionally, round bread stuffed with various ingredients is also a part of Turkish cuisine, such as the popular "ekmek dolması" (stuffed bread) commonly practiced in the Izmir province of Turkey, and "Fodula ekmek dolması."

Stuffing meat products is a common technique in Turkish cooking that infuses them with flavour and moisture while enhancing the texture and appearance of the final dish, particularly when cooked in a wood-fired oven. It is worth noting that while similar mixtures of ingredients listed in Table 6.6 can be used for filling, stuffing poultry and game typically does not involve minced or chopped meat products. However, it is possible to include pre-cooked meat products either outside or nested inside the main meat as well, allowing for fusion recipes.

When stuffing birds such as chicken or turkey, a variety of ingredients listed in Table 6.6 can be used depending on the desired flavour profile. However, it is crucial to ensure that the stuffing is cooked thoroughly and reaches a safe internal temperature to avoid the risk of foodborne illness. Therefore, a "slow cooking" method is recommended for stuffed meat products, except for seafood, which should be cooked more quickly. It is important to note that, unlike the current trend in Western cuisines, breadcrumb-based fillings are rarely used in traditional Turkish stuffed-meat dishes.

"Mussels" (Midye Dolma) is a highly popular delicacy in seaside cities of Turkey. The mussels are typically filled with a seasoned mixture of rice, onions, and herbs, then steamed or grilled. They are usually served with a squeeze of lemon and can be enjoyed as a snack or as a full meal.

In addition, an early form of sushi has been found in Turkish cuisine, with fillings placed inside deboned fish. A similar oven-cooked dish has also been discovered in one of the oldest recipes used in the palaces of the Ottoman Empire. Note that although deboning a fish is commonly used in other cultures, which is known as "butterflying," the method of deboning mentioned here is different, which keeps the fish as a whole suitable for stuffing.

Massaging the fish can help to loosen the bones and make them easier to remove, which is done as:

- Clean the fish, remove scales, gut it but keep the head and spine in place and rinse it thoroughly, and dry to remove excess moisture.
- Hold the fish firmly but gently by the head and tail, and apply gentle pressure along the length of the fish by using thumbs and gradually increase the pressure as needed. Start from the head and moving towards the tail, but focus on the area where the bones are located (usually along the centre of the fish following the line of the spine).
- The massaging action helps to soften the flesh and help to release the bones. Then loosen the head while keeping the spine in place and gently pull the head out that is attached to the loosened bone. Note that pulling the bones out gently in the same direction as the bone is critical.

Note that the effectiveness of this method can vary depending on the fish species and the size and condition of the bones. Exercise caution and be gentle to avoid damaging the flesh of the fish. Some fish species with larger bones and tender textures are suitable for this method including sea bass, red snapper, trout, branzino (European sea bass).

In addition, there are a few fish dishes in Turkish and Ottoman cuisine. One such dish is "İçli Köfte Balık," which originated from the south-eastern part of Turkey. In this dish, the fish is cleaned, seasoned with salt and pepper, and stuffed with a mixture of bulgur, ground meat, onions, and spices. The fish is then cooked in a pot with tomato sauce and served hot. Another fish dish is "Levrek Dolması," made with sea bass. In this dish, the fish is deboned and stuffed with a mixture of rice, pine nuts, currants, and spices. The fish is then baked in the oven with a tomato-based sauce. Additionally, Fish Lavangi, a dish with walnuts, is a popular dish in Azeri and Talesh cuisine, typically prepared during cooler weather. The recipe usually calls for white fish from the Caspian Sea.

As stated above, when using fish for stuffing in a recipe, it is recommended to use either deboned fish or a large fish with thick bones or small fish with edible bones. Three unique stuffed dishes in Turkish cuisine are illustrated in Figure 6.8.

FIGURE 6.8 Three of the selected stuffed dishes in Turkish cuisine are made of round breads, mussels and fish (wrapped in figs leaves).

Ertuğrul, Nesimi, Personal photograph, Adelaide, Australia, 2022.

Note that there are many variations of stuffed-bread, pastry, and pie dishes around the world. These dishes are typically filled with a savory or sweet filling and then baked until crispy on the outside and warm and gooey on the inside. Some popular examples include meat pies in Australia, Stromboli bread rolls in Italy and America, Calzone in Italy (which is essentially a folded pizza), Empanadas in Latin America, Pirozhki in Russia, and Bierocks in Germany.

6.4 MEATBALLS (KÖFTE)

The term "köfte" refers to a type of meatball dish that is widely enjoyed in Turkish cuisine, as well as in the Middle East, Central Asia, and the Balkans. The word "köfte" is believed to have originated from the Persian word "kufteh," meaning "minced/ground meat." This flavourful dish likely made its way to the Ottoman Empire through Persian or Arab traders, eventually becoming a beloved staple in Turkish cuisine.

Köfte is typically prepared using minced beef or lamb, or a combination of both. The meat is mixed with ingredients such as bread crumbs, bulgur (cracked wheat), garlic, onions, and a blend of spices like cumin and paprika. The mixture is then shaped into small balls or patties, which can be grilled, fried, or baked. The tempting aroma and smoky flavours that arise from grilling köfte in a wood-fired oven are truly exceptional. Additionally, there are popular vegetarian versions of köfte, including bulgur köfte and mercimekli (lentil) köfte, which do not require wood-fired oven cooking (although they may be boiled in water or fried in a pan) and serve as popular side dishes.

While the specific preparation and ingredients may vary, köfte is enjoyed in various forms such as İçli Köfte (stuffed köfte), Çiğ Köfte (raw köfte), and Adana Köfte (a spicy version from the Adana region of Türkiye). It is also a popular street food across Türkiye. Köfte is commonly served alongside rice or accompanied by an array of side dishes and condiments, including salad, yoghurt, and pickles. It can be wrapped in flatbread or served in a pita or pide bread. Numerous regional variations of köfte exist, each boasting its own distinct flavour profile and preparation method. However, a desirable texture for a cooked köfte is a crisp exterior with a tender and juicy interior.

It is important to note that meatballs come in various shapes, sizes, and local variations, all made from minced meat (often a combination of veal and lamb, with a desirable amount of fat). Onion is an essential ingredient, finely grated or chopped. Köfte is traditionally cooked on grills or hot plates, although using a wood-fired oven is a popular choice, as it imparts exceptional aroma and smokiness to the dish. Table 6.8 provides an overview of popular köfte types commonly found in Turkish cuisine, highlighting their specific ingredients and shapes. Common meatball shapes include disks, balls, oval balls, and pressed ovals, all traditionally formed by hand.

6.5 SHISH KEBABS (ŞİŞ KEBAB)

The word "kebab" is derived from the Persian word "kabāb." The exact origin of kabāb is not known, but it is believed to have originated in the Middle East or Central Asia, possibly in Persia or Anatolia (Asia Minor). Historical records suggest that kabāb was already a popular food in Persia during the time of the Achaemenid Empire (550–330 BC), and it is mentioned in the famous Persian epic poem Shahnameh (known as the "Book of Kings," written by the Persian poet Ferdowsi between 977–1010). Kebab likely evolved from earlier forms of grilled or roasted meat, which have been part of the region's cuisine for thousands of years. One of the most famous dishes of the Achaemenid Empire was called "chelow

TABLE 6.8 Ingredients used and types of meatballs (köfte) in Turkish Cuisine with regional names and features that can be cooked in wood fired ovens on grills or on trays, and serving options

MEATBALL Ingredients and Popular Types	Region/Shapes	
Commonly used Ingredients:		
The ingredients vary depending on the recipe of a specific region and unique shapes and texture. Common ingredients are: Minced beef, veal, lamb or chicken, fine bulgur, semolina, breadcrumbs, garlic, coriander, onion, tomato, tomato and red pepper pastes, herbs (dried mint, parsley, thyme, oregano, fresh mint leaves), shelled pistachios, walnuts, pine nuts, egg, spices (such as cumin, paprika, black pepper, red pepper flakes, chili pepper, mint, allspice, starch, curry powder), biscuit or cookie flour, baking soda, potato, water or meat stock, cheddar cheese, butter, olive oil, vegetable oil and salt.		
Turkish Names (with basic ingredients)		
Akçaabat Köfte (Ground beef:70% lean+30% fat), onion, fine bulgur, breadcrumbs, tomato paste, red pepper flakes, ground cumin, salt, black pepper)	Originated in the town of Akçaabat in Türkiye.	
Balaban Köfte (Ground lamb or beef, onion, garlic, breadcrumbs, egg, paprika, ground cumin, ground coriander, salt, pepper and olive oil).	Originated in the town of Eskisehir in Turkiye	
Cekirdekli Köfte (Ground beef, fine bulgur, onion, tomato and red pepper paste, dried mint, ground cumin, paprika, black pepper, salt, parsley, onion, shelled pistachios and olive oil)	Known as "Antep-style", which is a town named Gaziantep.	
Düzce Köfte (Ground beef, onion, egg, breadcrumbs, tomato paste, red pepper flakes, dried mint, salt and pepper.)	Originated in the town of Düzce.	
İçli Köfte (Shell made of fine bulgur with water and with ground meat, walnuts, pine nuts, onion, tomato, parsley, cumin, paprika and chili powder).	Originated in the town of Şanlıurfa	
İnegöl Köfte (Ground beef with 20% fat, onion, egg, breadcrumbs or stale bread, salt, black pepper, cumin, thyme, oregano and vegetable oil.)	Originated in the town of İnegöl/Bursa.	
Turkish Names (with basic ingredients)		
Islama Köfte (Ground beef, egg, onion, garlic, cookie flour, fine bulgur, parsley, breadcrambs, butter, chilie pepper, water or meat stock, salt, black pepper, oregano and cumin.)	Originated in the town of Adapazari,	
İzmir Köfte (Ground beef with 20% fat, onion, breadcrumbs, egg, salt, black pepper, cumin, paprika, red pepper flakes, parsley and fresh mint leaves).	Originated in the town of İzmir.	

(Continued)

TABLE 6.8 *(Continued)* Ingredients used and types of meatballs (köfte) in Turkish Cuisine with regional names and features that can be cooked in wood fired ovens on grills or on trays, and serving options

Kadınbudu Köfte (Pre-cooked lamb or beef mince and rice, onion, garlic, herbs, dill, mint, salt, black pepper, paprika, cumin, tomato paste and vegetable oil for frying. Formed to a ball shape and covered by flour/egg mixture before frying.	Popular dish during Ottoman time.	
Kasarli Köfte Minced veal and lamb, onion, cookie flour, egg, parsley, cheddar cheese (to be placed in the centre of the ball shape meatballs), oregano, black pepper, cumin, salt and butter (for brushing before cooking in the oven).	Originated in the town of Tekirdag.	
Kıbrıs Köfte (Minced beef or lamb, fine bulgur, onion, garlic, tomato paste, red pepper paste, cumin, dried mint, dried oregano, salt and pepper and olive oil).	Originated in Cypress.	
Kuru Köfte (Minced beef or lamb, fine bulgur or breadcrumbs, garlic, onion, tomato paste, parsley, cumin, salt, black pepper, red pepper flakes, paprika, salt and olive oil).	Originated in the town of Kütahya	
Misket Köfte (Minced beef or lamb, fine bulgur, garlic, onion, tomato paste, cumin, salt, dried mint, red pepper flakes, salt and olive oil).	Originated in the town of Tekirdag.	
Tavuk Köfte (Minced chicken, onion, egg, olive oil, allspice, starch, curry powder, cookie flour, parsley, black pepper, cumin, salt and oregano).	Originated in the town of Çorum	

Turkish Names (with basic ingredients)

Tekirdag Köfte (Minced veal/lamb with medium fat content, onion, garlic, bread crumbs or fine semolina, baking soda, black pepper, chili pepper, cumin, salt).	Originated in the town of Tekirdag.	
Urgup Köfte (Potato, minced veal or lamb, onion, garlic, fine bulgur, salt, black pepper, allspice, tomato paste, red pepper flakes, paprika cumin, salt and olive oil)	Originated in the town of Urgup.	

KOFTE SERVING OPTIONS

The köfte is typically accompanied by a variety of ingredients that complement its flavours. Some common serving options include:

- Serving on a pide or a yufka bread.
- Sliced fresh tomatoes and green peppers, whether sweet or hot.
- Gently roasted or grilled tomatoes, peppers, and/or onions.
- Fresh parsley leaves.
- Lettuce and pickles.
- Tomato sauce made with tomato paste, butter, red pepper flakes, and salt (for Islama köfte).
- Tomato sauce made with olive oil, chopped onions, chopped garlic, tomatoes, tomato paste, salt, sugar, black pepper, dried thyme, and water (for Izmir köfte).

TABLE 6.8 (Continued) Ingredients used and types of meatballs (köfte) in Turkish Cuisine with regional names and features that can be cooked in wood fired ovens on grills or on trays, and serving options

- Sweet sauce made with yogurt, finely chopped dill, mint, olive oil, and salt.
- Hot sauce (cooked) made with grated tomatoes, tomato paste, pepper paste, hot pepper sauce, onions, garlic, olive oil, oregano, basil, and salt.
- Piyaz, a white bean salad made with cooked small beans, finely chopped onions, chopped spring onions, chopped tomatoes (cubed), finely chopped parsley, lemon juice, olive oil, salt, black pepper, chili pepper, sumac, and chopped hard-boiled egg.
- The combination of grilled köfte and piyaz has become a delicious fast-food option in Türkiye.
- A whole loaf of bread cut in half and packed with grilled meatballs, lettuce, tomatoes, and shredded onions.
- Cooked bulgur with chopped onions and sumac served with pide and yufka bread.
- Serving köfte on the side or on top of cooked rice or bulgur.
- Cooking köfte in casseroles for a different presentation.

kabab," which is a type of kebab made with marinated and grilled meat served over a bed of saffron rice. This dish remains popular in modern-day Iran and is considered a national dish.

The history of kebab is diverse and complex, with many different types of dishes and cooking methods that have evolved over time in different parts of the world. Today, kebab is a popular dish enjoyed in many countries around the world, including Türkiye, the Middle East, Iran, Central Asia, Turkic countries and India. The type of meat used for kebab varies depending on the region and can include beef, lamb, chicken, fish, and vegetables.

Although the evolution of kebab dishes over time suggests that grilled meat may have led to the development of the "skewered" version known as "shish kebab," there is no written evidence to support this theory.

The name "şiş kebab" comes from the Turkish word "şiş," which means "skewer." Shish kebab is a popular Turkish dish enjoyed worldwide. It consists of marinated cubes of meat that are skewered and grilled. Shish kebab has many regional variations, such as "Joojeh kebab" in Iran, which is made with marinated young chicken, and "Lahm Meshwi" in Lebanon, made with lamb or beef and marinated in a mixture of olive oil, garlic, lemon juice, cumin, coriander, and paprika. Lahm Meshwi is usually served with a variety of mezes, including hummus, tabbouleh, fattoush, and pita bread.

Table 6.9 summarizes the most well-known types of Turkish shish kebabs, each with its own distinct flavours and preparation methods. In addition to cubed meat-based kebabs, there are also minced meat kebabs formed into long meatballs on a skewer, as well as kebabs made entirely with sliced vegetables or a mixture of cubed meat and vegetables on the same skewer.

To prepare shish kebab, the meat is typically marinated in a blend of spices, herbs, and olive oil for several hours or even overnight to enhance its flavour and tenderness. After marinating, the meat is skewered and cooked over a wood-fired oven until fully cooked and slightly charred on the outside. Shish kebab is often served with a variety of side dishes and condiments, such as rice pilaf, grilled vegetables, salad, and yoghurt sauce.

6.6 KEBABS/GUVEC (CLAYPOT)/CASSEROLES

The origin of Güveç (clay pot) is not clear, as the dish has been a part of Turkish cuisine for centuries and has likely evolved over time with various cultural and regional influences. Clay pots are known for their

TABLE 6.9   Well-known "**Shish Kebabs**" in Türkiye, suitable for wood-fired oven cooking

Cubed-Meat Style Shish Kebabs

Ciğer Şiş (Liver Shish): Marinated lamb or beef liver, and seasoned.

Balık Şiş (Fish Shish): Marinated fish.

Çöp Şiş (Tiny/Small shish): Marinated lamb or beef, seasoned with salt, pepper, paprika.

Dana Şiş (Veal Shish): Marinated veal tenderloin.

Kalamar Şiş (Calamari Shish): Calamari is marinated with olive oil, lemon juice, garlic, and herbs.

Kanat Şiş (Wing Shish): Marinated chicken wings.

Kanlıca Şiş: Marinated lamb or beef served with yoghurt sauce.

Karışık Şiş (Mixed Shish): With a variety of marinated lamb, beef, chicken, and sometimes liver.

Kılıçbalığı kebabı (Swordfish Shish Kebab): Marinated swordfish.

Kuzu Şiş (Lamb Shish): Marinated lamb, sometimes served with grilled vegetables.

Piliç/Tavuk Şiş (Chicken Shish): Marinated chicken breast or thigh.

Soslu Şiş (Shish with Sauce): Marinated lamb, beef or chicken served with a tomato-based sauce.

Tereyağlı Şiş (Shish with Butter): Marinated lamb or beef served with melted butter on top.

Kaşarlı Şiş (Cheddar Shish): Marinated lamb or beef cooked with cheddar cheese.

Midye Kebabı (Mussels Shish Kebab): Mussels, marinated with olive oil, lemon juice and herbs.

Kuzu Tandır (Slow-cooked Lamb Shish): Slow-cooked lamb served on a skewer.

Meatball Style, longitudinally formed on skewers

Adana kebab (Adana Shish Kebab): Traditionally with minced lamb, occasionally with minced beef, prepared with red pepper flakes or Aleppo pepper, and with the choices of cumin, sumac, salt, black pepper and onion, all shaped onto long, flat metal skewers. Often served with rice, salad, and grilled vegetables. Named after the city of Adana in Türkiye.

İzmir Köfte Kebabı (İzmir Shish Kebab): meatballs are flavoured with garlic and herbs, and loaded on a skewer. A variant of Adana or Urfa Kebab!

Kıyma Şiş/Kıyma Kebabı (Minced Shish Kebab) or Köfte Şiş (Meatball Shish): Minced lamb or beef or combination that is mixed with spices, shaped into sausage-like form. A variant of Adana or Urfa Kebab.

Kokoreç (Roasted Intestine): A type of offal dish cooked on a skewer also exist that can be classified under this group. The inner part sweetbreads is wrapped around a skewer first with onions and spices, and then the thin lamb intestine is wrapped around it with thin layers of animal fat. Served with finely chopped tomato, oregano and chilli pepper after roasting.

Şiş Köfte (Shish Meatball): Minced lamb or beef, mixed with bulgur, onion, and spices. The mixture is formed into small balls or patties and then loaded on skewers.

Urfa Kebabı (Urfa Shish Kebab): Minced beef or lamb with isot pepper (also known as Urfa pepper, a type of dried chilli pepper that is native to the region), paprika, cumin, sumac, garlic, parsley and salt, then shaped into elongated patties on skewers, resulting slightly spicy kebab often served with rice or bread and a side of yoghurt sauce. Named after the city of Urfa in Türkiye.

Mixed Types: Cubed-Meat + Vegetables

Çoban Kebabı (Shepard Kebab): Lamb or beef with tomatoes, peppers, and onions.

Külbastı (Coal-Roasted or Ember-Roasted Meat): Marinated lamb or beef with grilled tomatoes and peppers.

Mantar Kebabı (Mushroom Shish Kebab): Marinated mushrooms plain or with lamb or beef.

Şeftali Kebabı (Peach Shish Kebab): Minced lamb or beef seasoned with onion, parsley and spices, shaped into small patties around a skewer, and served with grilled peaches.

Soğan Kebabı (Onion Shish Kebab): Marinated lamb or beef with onions.

Tokat Kebab: Cubed and marinated (with olive oil, garlic, onions, thyme, cumin, and black pepper) lamb or beef with eggplants, tomatoes, peppers, and potatoes are all grilled together on a single skewer. It is originated from Tokat province in Türkiye.

ability to retain heat and moisture, making them ideal for slow-cooking stews and casseroles. Although kebabs, güveç (clay pot), and casseroles may be considered different types of dishes, they share many similarities, including:

- All three types of dishes can be made with beef, lamb, and chicken, and commonly use vegetables, making them "one-pot cooking" options suitable for feeding a crowd or for meal prep.
- All three dishes are ideally cooked slowly, which makes them a perfect match for wood-fired oven cooking. Slow cooking allows the flavours to develop and the ingredients to become tender.
- All three dishes incorporate a variety of spices and herbs to add flavour. Kebabs often include a spice mixture, while güveç and casseroles often feature herbs like thyme and rosemary. However, the seasoning can be customized based on individual preferences. For example, vegetarian versions of güveç are popular, including ingredients like eggplant, green beans, potatoes, and onions. Casseroles may include pasta or rice.
- All three dishes are cooked inside a suitable pot with a lid, but traditionally güveç is cooked in a clay pot. Güveç typically consists of meat (lamb, beef, or goat), vegetables (such as eggplant, potatoes, tomatoes, and green beans), and a variety of spices. Kebabs can be made with meat, vegetables, and various fruits (such as dates, prunes, apricots, or chestnuts). Casseroles typically include a mixture of ingredients (meat, vegetables, pasta, or rice) along with a sauce or cheese.
- They can all be served as main dishes and are often accompanied by rice, bulgur pilaf, and plain pide or other types of bread.
- While kebabs are often associated with Middle Eastern and South Asian cuisine, güveç is a popular dish in Turkish cuisine, and casseroles can be found in many Western and Mediterranean cuisines.

In Türkiye, the "Kebab" family of dishes using "Güveç" has a wide variety of regional and ingredient-based variations, which are summarized in Table 6.10. Most of the kebabs have unique names associated with their specific region and culinary traditions. All three groups of dishes are often served with a side of rice or bulgur pilaf, along with fresh salads, yoghurt dips, or grilled vegetables. The meats used in these dishes (lamb, beef, chicken, goat) are typically seasoned with spices such as cumin, paprika, and garlic, often marinated overnight to enhance the flavour.

It is important to note that the term "Kebab" has become synonymous with two well-known dishes in Turkey: "Döner Kebab" and "Şiş Kebab." In Turkey, "Döner Kebab" is available in two different orientations: vertical standing and horizontal standing, known as "Cag döner," which is suitable for wood-fired oven cooking. It is worth mentioning that the style of "Döner cooking" is not exclusive to Turkish cuisine and can be found in various culinary traditions worldwide. The exact origins of this cooking method are debated, but it is likely influenced by the migration and trade routes of different peoples throughout history. Similar dishes can be found in other cultures, such as "Yakitori" in Japan, "Brochette" in France, "Espetada" in Portugal, "Satay" in Malaysia, Thailand, Singapore, and the Philippines, "Chuan" in China, "Suvlaki" in Greece, "Mtsvadi" in Georgia, and "Shashlik" in Crimea, Kazakhstan, Lithuania, Poland, Ukraine, and Russia.

6.7 LEGUMES IN CLAY POTS WITH LIDS

Cooking legumes in a clay pot is a traditional method that has been practiced for centuries across various cultures. Clay pots are favoured for cooking legumes because they retain heat and moisture exceptionally

TABLE 6.10   Well-known kebabs and guvec (claypot)/casseroles in Türkiye

Kebabs

Abant Kebab: Slow-cooked lamb or beef (marinated by cumin, paprika, and red pepper flakes) stew with potatoes, carrots and onions.

Ali Nazik Kebabı: Made with grilled or roasted aubergine (eggplant) puree mixed with yoghurt, and topped with seasoned cooked lamb or beef.

Beyti Kebab: It is a pastry dish originated in Istanbul. A meatball shish is made with minced lamb or beef, and then shaped into long cylindrical meatballs. The meat is then wrapped in lavash or thin flatbread and grilled until cooked through. After cooking, the meat is sliced and served with yoghurt sauce, tomato sauce, and sometimes topped with melted butter. It is often served with grilled vegetables and rice pilaf on the side.

Buryan Kebab: Slow-cooked lamb or chicken (marinated with cumin, coriander, cinnamon, and cardamom) layered with fragrant basmati rice (that is cooked with saffron, cinnamon and cloves). The dish is assembled by layering the meat and rice in a large dish, which is then sealed with bread dough and baked in wood-fired oven for several hours. The meat is served on a bed of rice, garnished with nuts, raisins, and herbs.

Çökertme Kebabı: Marinated (with paprika, cumin, and oregano) lamb or beef is cooked and thinly sliced (as in doner kebab) are served on a bed of pita bread and topped with tomato sauce, yoghurt and melted butter. Served over a bed of grilled tomatoes, peppers and onions and pide.
cooked in a sealed parchment (baking) paper with vegetables (such as potatoes, tomatoes, onions, and peppers) and spices along with some olive oil and salt. It can be served directly from the paper and with rice or pide bread and is often accompanied by a side salad or yogurt sauce.

Etli yaprak sarma kebabı (Wrapped grape leaves with minced meat): This is a mixed-style dish! A mixture of pre-cooked rice and meat, onions, tomatoes, herbs, and spices are wrapped with tender grape leaves and cooked. Then meat (beef or lamb) is marinated with cumin, paprika, and garlic, then cooked as a shish kebab and served on top of wrapped grape leaves dish.

Güveç kebabı: Consists of lamb or beef, and vegetables, such as eggplant, peppers, potatoes, onions, and tomatoes. The meat is marinated in paprika, cumin, and garlic. The clay pot is sealed with a lid and cooked in the oven. It is often served with rice or bulgur pilaf.

İskender kebabı: This is a highly popular kebab originated in the province of Bursa. The recipe is very similar to Çökertme Kebabı, and it may be served with a dollop of yoghurt on the side and served with a fresh green salad.

Islim Kebab (Thin-Slice Kebab): This traditional Turkish dish consists of grilled lamb or beef meat inside the thin slices of eggplant. The marinated and cooked meat is placed onto the eggplant slices and rolled up tightly, and cooked again on a skewer. It is served hot with a side of yoghurt sauce or tomato-based sauce and rice or pide bread

Kağıt Kebabı (Paper Kebab): Lamb or beef is marinated with paprika, cumin, and garlic and

Kilis Tava Kebab: Lamb or beef marinated with paprika, cumin, oregano, and garlic and is pan-fried with onions until browned on all sides and sliced tomatoes, green peppers, and hot peppers are added to the pan and cooked, and served in a traditional copper or iron skillet called a "tava." It is originated in the town named Kilis in Türkiye.

Macar Kebab (Hungarian Kebab): Made with minced lamb or beef, which is seasoned with cumin and paprika. The meat is then shaped into small meatballs or patties and grilled, and served on a bed of flatbread, along with a generous amount of melted cheese and a variety of grilled vegetables such as onions, tomatoes, and peppers. The dish is often topped with a dollop of yoghurt or a spicy tomato sauce.

Manisa Bohca Kebab (Manisa Pouch Kebab): Marinated lamb meat is skewered and grilled, and then wrapped in a thin layer of dough to form a pouch around the meat and baked until crispy and golden brown. It is served hot with rice, salad, or grilled vegetables.

TABLE 6.10 (Continued)   Well-known kebabs and guvec (claypot)/casseroles in Türkiye

Manisa Kebab/Şehzade Kebabı: Marinated lamb with garlic, paprika, and cumin is skewered and grilled, and is served on a bed of pide bread or lavash bread, which is grilled again to make it crispy, and is garnished with grilled tomatoes, green peppers, and onions.

Orman Kebab (Forest Kebab)/Testi Kebab: Cubed lamb, beef, veal, goat, chicken meat, onions, tomatoes and green peppers with various spices and herbs (paprika, cumin, and parsley) all cooked together inside an earthenware pot ("testi" in Turkish) after sealing with a bread dough or aluminium foil. It is at the table by braking the seal, usually with pide bread.

Oruk Kebab/Kibbeh/Icli Kofte: These three different named dishes are all oval bulgur balls with various fillings. To make oval balls, the bulgur dough is made by mixing fine bulgur with wheat flour, tomato paste, red pepper paste, cumin, salt and black pepper, and in some versions egg and mashed potato. Pre-cooked filling can be made using a combination of minced beef, onion, walnut, pistachios, salt, ground black pepper, paprika, red pepper flakes, and lentils in vegetarian version. Then the mixture is shaped into small oval balls with filling. Kibbeh (Lebanese dish) uses minced meat, allspice and pine nuts as filling and the bulgur balls are fried until lightly browned. Oruk Kebab and Icli Kofte use minced meat, walnuts and pistachios as filling. However, Oruk Kebab is baked until lightly browned and Icli Kofte is first boiled in water until firm then sautéed in tomato sauce, and both can be served with melted butter. All are known for their nutty, slightly spicy flavour and crunchy texture on the outside with a soft, chewy interior.

Patates Kebabı (Potato Kebab): Sliced potatoes are cooked with onions, peppers and tomatoes in a rich tomato-based sauce (with cumin, paprika and oregano), and sometimes, eggplant or zucchini is included. The vegetables are layered in a baking dish and topped with a layer of mozzarella or feta cheese, and baked.

Patlıcanlı Kebabı (Eggplant Kebab): The eggplants are sliced lengthwise and either roasted or grilled, while the meat is marinated with garlic, paprika, cumin, and pepper and grilled or cooked in a pan. The cooked eggplants are layered with the cooked meat, and the dish is served with a tomato-based sauce, yogurt, and sometimes rice or bulgur pilaf.

Sulu köfte kebabı (Watery Meatball Kebab): The round shape meatballs are made from minced beef or lamb, rice, grated onions, cumin, paprika and black pepper, and are browned in a pan or on a grill. The tomato-based sauce is prepared from onions, garlic, tomatoes, red pepper flakes, cumin, and coriander, and the meatballs are added together with water or broth to simmer until cooked.

Sütlü Kebabı (Milky Kebab): Marinated lamb or beef in a mixture of olive oil, garlic, cumin, paprika, and black pepper is skewered and grilled, and is covered by milk-based sauce that is made with a mixture of milk, flour, butter, salt and black pepper. It is served with a side of rice or bulgur pilaf and a fresh salad or grilled vegetables.

Tas Kebab (Bowl Kebab): Lamb or beef is browned in oil with onions, garlic, paprika, cumin, and black pepper. Tomatoes, bell peppers, and sometimes potatoes or carrots are added, and it is slowly simmered. Served with rice or bulgur pilaf and accompanied by yoghurt or a side salad.

Tepsi Kebab (Tray Kebab): Seasoned minced lamb or beef is formed into small patties or meatballs with onions, garlic, paprika, cumin, and black pepper. It is arranged in a tray with sliced vegetables like tomatoes, bell peppers, and sometimes eggplants or potatoes, and baked in the oven. Served hot with rice or bulgur pilaf, a side salad, and yoghurt or tangy tomato sauce.

Yalancı Kebab (Fake Kebab): Bulgur is soaked in hot water and mixed with onions, tomatoes, peppers, parsley, mint, and paprika. The mixture is shaped into small oblong patties or balls and grilled with eggplant. Served hot with yoghurt, tomato sauce, or spicy pepper paste. Can be accompanied by pide bread or rice pilaf.

Guvec/Casseroles

Çoban Kavurma Güveç (Shepherd's Stir-Fry Clay Pot): Cubed lamb (or beef) meat is sautéed in a pan with onions and garlic until browned. Sliced bell peppers and tomatoes are added with paprika, cumin and oregano, and transferred to a clay pot and cooked slowly.

Etli Güveç (Meaty Clay Pot Stew): This is a classic Turkish güveç dish. Cubed lamb or beef are placed at the bottom of the güveç, followed by available vegetables, potatoes, onions, garlic, bell peppers, eggplants and zucchini. It is seasoned with black pepper, paprika, cumin, garlic and a touch of tomato paste and a small

(Continued)

TABLE 6.10 (Continued)   Well-known kebabs and guvec (claypot)/casseroles in Türkiye

amount of water or broth and olive oil is added and the lid is covered and slow-cooked for several hours. It is served with rice or bulgur pilaf and crusty bread or a side salad or yogurt.

Güveçte Pilav (Rice Pilaf in Clay Pot): Rice is washed and soaked and onions, tomatoes, and peppers are chopped and sautéed in a pan with cumin, paprika, and parsley. The mixture is slow cooked in a clay pot with covered with a lid that is fully sealed. It is served hot with a side of yoghurt or pickled vegetables, and is garnished with fresh parsley or cilantro.

Hamsili Güveç (Anchovy Clay Pot): Fresh anchovies are cleaned, gutted, and deboned if larger, and they are layered in the clay pot, alternating with layers of sliced onions, tomatoes, and green peppers, and/or zucchini, eggplant and potatoes. The dish is seasoned with black pepper, crushed red pepper flakes, and paprika, and with fresh herbs. Then olive oil, water or fish stock are added to keep the fish moist during cooking that is performed when the lid is closed at a moderate temperature.

Imam Bayildi Güveç (Eggplant Clay Pot Stew): Stuffing in this dish is made of tomatoes, bell peppers, onions, garlic, and sometimes pine nuts and currants, and with a mix of black pepper, paprika, thyme, parsley or mint. The eggplants are hollowed out or scored, and the stuffing mixture is packed into the cavities or layered between the eggplant slices. Olive oil and a small amount of water or tomato juice are poured over the eggplants and slow cooked in the clay pot with a lid.

Kabaklı Güveç (Zucchini Casserole): This güveç dish is made with zucchini, minced beef, onions, tomatoes, paprika, cinnamon, cumin, parsley and allspice, and cooked similar to the dishes described above.
Karidesli Güveç (Shrimp Clay Pot Stew): This is a traditional Turkish seafood güveç. Fresh, peeled shrimp with tomatoes, bell peppers, onions, garlic, and sometimes zucchini or mushrooms are seasoned with black pepper, paprika and thyme. Olive oil, lemon juice, and a small amount of water is added and slow-cooked at a low temperature.

Guvec/Casseroles

Kuru Fasulyeli Güveç (Dry Bean Clay Pot Stew): Pre-soaked white dry beans are slow cooked with diced tomatoes, chopped bell peppers, onions, garlic and seasoned with black pepper, paprika, thyme, salt and tomato paste. Olive oil and a small amount of water or vegetable broth are also added over the ingredients.

Kuzu Güveç (Lamb Casserole): Cubed lamb with tomatoes, bell peppers, onions, garlic, eggplant or zucchini are seasoned with black pepper, paprika, and thyme and fresh parsley in a clay pot. Water, tomato juice and/ or broth is added over the mixture and slow cooked. The dish is served with rice or bulgur pilaf and crusty bread.

Mantarlı Güveç (Mushroom Casserole): Mushrooms are cleaned, sliced, or quartered, and sautéed in olive oil or butter, followed by the similar ingredients and steps as above.

Patlıcanlı Güveç (Eggplant Clay Pot): Eggplant is peeled, sliced or cubed, and then roasted until slightly softened, and onions, garlic, tomatoes, bell peppers and zucchini are added with black pepper, paprika, cumin, oregano, parsley, mint, or dill. The ingredients are combined with water, tomato juice and/or vegetable broth and then slow cooked.

Sebzeli Güveç (Vegetable Clay Pot Stew): This is a vegetarian slow-cooked güveç very similar to the dishes described above. However, a variety of vegetables vary depending on the season and availability including green flat beans.

Sütlü Kabak Güveç (Creamy Zucchini Clay Pot Stew): Sliced zucchini with tomatoes, bell peppers, onions, and garlic are seasoned with black pepper, paprika, dill, fresh parsley or mint. A creamy sauce made from milk or yoghurt and a touch of flour is prepared and poured over to cover the mixture in the clay pot and slow-cooked.

Tavuklu Güveç (Chicken Clay Pot Stew): It is similar to Etli Güveç described above with the meat is replaced by chicken pieces, usually bone-in and skinless, and with the addition of sliced tomatoes and mushrooms.

TABLE 6.10 (Continued)   Well-known kebabs and guvec (claypot)/casseroles in Türkiye

Others

Döner Kebab:

Döner kebab is made from stacked layers of meat (beef, lamb and chicken) that are roasted on a spit and then shaved off as they cook.

Cooking on Kremit

There are various other kebab versions, with similar ingredients, but cooked in a shallow clay pot (also known as clay dish, clay saucer), which is called "kiremit" (can be translated to English as roof tile), which is also used as a serving plate. Some known popular names of "kiremit" dishes that are grilled or baked in the wood-fired oven are:

Kiremit Goat: Similar to Kiremit Kuzu, this dish uses goat meat instead of lamb.

Kiremit Kuzu: Marinated young-lamb pieces are cubed or sliced into thin strips, then seasoned with traditional Turkish spices are cooked, and is served with sides such as rice, vegetables, or salad.

Kiremitte Balık: A dish with whole fish or fish fillets marinated with herbs and spices.

Kiremitte Börek: A savoury pastry filled with cheese, spinach, or ground meat, baked until crispy and golden.

Kiremitte Hellim: Grilled or baked halloumi cheese, a popular meze in Türkiye.

Kiremitte Kaşarlı Tavuk Sote: This dish consists of diced chicken pieces sautéed with bell peppers, onions, and tomatoes, and topped with cheddar cheese, a semi-hard Turkish cheese, and grilled.

Kiremitte Kavurma Et: Features various cuts of meat, often beef or lamb, marinated with herbs and spices. The meat is typically sliced into thin pieces for even cooking, and is often served with a side of rice, vegetables, or salad.

Cooking on Kremit

Kiremitte Köfte: Seasoned meatball dish, often served with a side of rice or salad.

Kiremitte Mantar: Marinated mushroom dish, usually served as a side dish or appetizer.

Kiremitte Pide: A traditional Turkish flatbread similar to pizza, topped with various ingredients like cheese, ground meat, or vegetables, and baked on the kiremit.

Kiremitte Sebzeli Kebap: This is a vegetable kebab with eggplants, bell peppers, onions, and tomatoes.

Kiremitte Sucuk: A popular Turkish sausage "sucuk" dish, served as a meze or snack.

Kiremitte Tavuk: Marinated chicken pieces, typically served with a side of vegetables or rice.

Roastings in Sac (a convex metal griddle)

resemble shallow wok cooking: The sac is a large, shallow, dome-shaped pan that can be used on both sides. The convex side is used for bread products like "katmer," while the concave side is used for roasting and fast cooking various mixtures. This includes chopped vegetables, meat (such as lamb and beef), and re-roasting "kokoreç" with spice mix. Here are the dishes:

Dana Sac Kavurma (Veal Roasting in Sac): Diced beef is sautéed with onions, bell peppers, tomatoes, and spices like black pepper, red pepper flakes, and thyme. Cooked over high heat for a rich, caramelized flavour. Served with rice, bulgur, or bread.

Tavuk Sac Kavurma (Chicken Roasting in Sac): Similar to Dana Sac Kavurma, but with diced chicken. Sautéed with onions, bell peppers, tomatoes, and spices, then cooked over high heat in the sac. Served with rice, bulgur, or bread.

Kuzu Sac Kavurma (Lamb Roasting in Sac): Diced lamb sautéed with onions, bell peppers, tomatoes, and spices. Cooked over high heat in the sac for a rich, caramelized flavour. Served with rice, bulgur, or bread.

(Continued)

TABLE 6.10 *(Continued)* Well-known kebabs and guvec (claypot)/casseroles in Türkiye

Kokoreç: A popular Turkish street food made from lamb or goat intestines. Skewered with seasoned offal and grilled or roasted for a smoky flavour. Finely chopped and mixed with spices in a sac, then gently roasted. Served in warm loaf bread or on a plate with vegetables, pickles, and lemon wedges. Can also be served with rice, bulgur, or pilaf. Known for its rich, savoury taste and unique texture, enjoyed as a late-night snack or quick meal.

Fish in a Salt Crust:

Fish cooked in a salt crust is a highly popular cooking technique in Türkiye as that seals in moisture while enhancing its natural flavours. A brief description of the process:
- Commonly used fish includes sea bass and red snapper. The head and tail of the properly cleaned fish is left intact, and it is lightly seasoned with herbs and lemon slices.
- The salt crust mixture is prepared by coarse sea salt and a small amount of water or egg whites until a slightly damp, sand-like consistency is achieved.
- A layer of the salt mixture is spread on the bottom of a stainless-steel or earthenware baking pan, and the seasoned fish goes on top and then is covered completely with the salt mixture to create a tight seal.
- When the temperature of the wood-fired oven is around 200–230°C, the fish is cooked. A general rule is to cook the fish for 10 minutes per 2.5 cm of thickness.
- After the fish has finished cooking, it should be rested for few minutes before using a knife or mallet to crack the salt crust and carefully removing the hardened crust, taking care not to damage the fish underneath. The fish is served with a choice of sides or sauces.

well, resulting in even cooking and preserving the natural flavours and nutrients of the legumes. When cooked in a wood-fired oven, legumes acquire a unique and rustic smoky flavour that cannot be replicated in a regular oven. It is important to note the following considerations when cooking legumes in a clay pot:

- Soaking the legumes in cold water for several hours (ideally more than 4 hours) softens them and reduces cooking time.
- Prior to cooking, soak the clay pot and lid in water for approximately 2 hours. Add all the ingredients, followed by sufficient water or broth to cover them, ensuring the liquid level is about 1 cm below the pot's rim to prevent overflow while boiling.
- Slow cooking at lower temperatures with the oven door closed is generally preferred. However, if opting for open-door cooking, temperatures around 150–175°C and placing the clay pot in "Area E" (refer to the Data Logging Chapter) are recommended.
- If using a new clay pot, it may require "seasoning." This involves immersing the pot in water for a few hours or overnight, allowing it to dry, then rubbing the interior with vegetable oil or butter. Finally, bake the pot in the oven at a low temperature for approximately an hour. This process strengthens the pot and improves its durability. Note that the quality of the pot and glaze can influence the seasoning process. After cooking, clean the clay pot with warm water and a soft sponge, avoiding abrasive cleaners and brushes.

Common legumes used in Turkish cuisine include kidney beans, white beans, lima beans, fava beans, great northern beans, green/dry broad beans (Bakla), flat green beans, lentils, pinto/borlotti (or cranberry) beans, chickpeas, and yellow/brown/green/red lentils. Meat and meat products commonly used are beef, veal, lamb, goat, chicken, Turkish sucuk, and pastırma. Many traditional Turkish dishes, including those with legumes, can be adapted to be cooked in a güveç (clay pot). Remember to use lower temperatures and longer cooking times for best results. Refer to Table 6.11 for a list of Turkish dishes featuring legumes.

TABLE 6.11   Turkish dishes using legumes that can be prepared for clay pot cooking

LEGUMES IN CLAY-POT

It is recommended to use olive oil in legume cooking. The most popular legume dish in Turkish Cuisine is Dry Beans and their multiple versions in a clay pot. In addition, Turkish sucuk and pastirami are highly regarded ingredients in dry legumes.

Bamya (Okra in Clay Pot): A dish made with fresh okra, tomatoes, onions, and sometimes chickpeas, flavoured with black pepper, red pepper flakes, and dried mint. Sun-dried okra may also be used but after soaking in water prior to cooking.

Barbunya Pilaki (Pinto/ Borlotti Beans in Clay Pot): A dish made with Pinto/Borlotti beans, cooked with onions, tomatoes, and carrots, with black pepper, red pepper flakes, and a touch of sugar. It is typically served cold or at room temperature as a meze or appetizer

Etli Kuru Fasulye (Dry Beans with Meat in Clay Pot): This is a variation of Kuru Fasulye which includes meat such as beef, lamb, or goat.

Etli Taze Fasulye (Green Beans with Meat in Clay Pot): This is a variation of Etli Kuru Fasulye but with green bean.

Kuru Fasulye (Dry Beans in Clay Pot): A traditional Turkish white bean stew cooked with tomatoes, onions, and red or green peppers, flavoured with paprika, cumin, and red pepper flakes.

Mercimek Çorbası (Lentil Soup in Clay Pot): A hearty red lentil soup that includes onion, carrot, and tomato paste, and with cumin, paprika, and dried mint.

Mercimek Köftesi (Lentil Meatballs in Clay Pot): As it was mentioned previously, red lentil "meatballs" are made with bulgur wheat, onions, tomato paste, and with cumin, red pepper flakes, and black pepper. It is highly recommended to cook in a clay pot to infuse the dish with a smoky flavour.

Nohut Yemeği (Chickpeas Stew in Clay Pot): A chickpea stew made with onions, tomatoes, and green peppers and typically seasoned with cumin, black pepper, and red pepper flakes.

Soganli Kuru Fasulye (Dry Beans with Onions in Clay Pot): This version is very similar to the Kuru Fasulye, but the ratio of beans/onions should be 1 and no red or green peppers are used.

6.8 PASTRIES AND DESSERTS

In Turkish cuisine, a wide array of pastries and desserts can be prepared using wood-fired ovens. Table 6.12 provides a brief description of select pastries and desserts, categorized into Börek, Baklava, Kadayif, Puddings, Fruits, and Others. These choices have a longstanding tradition and enjoy popularity. It is important to note that Turkish cuisine offers numerous variations and regional specialities that also utilize this traditional cooking method.

6.8.1 Börek

Börek is a savoury pastry originating from the Ottoman Empire and now enjoyed as a popular snack and breakfast dish in the Balkans, Middle East, and Central Asia. It consists of layers of thin, flaky phyllo (yufka) pastry dough, filled with a variety of ingredients such as minced meat, cheese, spinach, or potatoes. Böreks can be baked in different shapes and sizes, including flat-form rolls, triangles, or coils. When baked in wood-fired ovens, börek develops a smoky flavour and crispy texture. It is important to closely monitor the oven temperature, aiming for around 170–200°C, as fluctuations may occur during the cooking process to achieve optimal results.

TABLE 6.12   Pastries and Desserts in Turkish Cuisines that can be cooked in wood-fired oven

Pastries
Börek:

Ispanaklı Börek (Spinach Börek): This is made with layers of phyllo dough filled with a mixture of spinach, onions, and cheese. Baking it in a wood-fired oven creates a crispy, smoky crust.

Kıymalı Börek (Minced Meat Börek): It features a filling of minced meat, onions, black pepper and paprika, sandwiched between layers of phyllo dough. Baking it in a wood-fired oven creates a deliciously crispy pastry.

Patatesli Börek (Potato Börek): It is filled with a mixture of mashed potatoes, onions, and spices, such as black pepper and red pepper flakes. Baking it in a wood-fired oven adds a smoky flavour and a crispy crust.

Sigara Böreği (Cigar-Shaped Cheese Börek): Cigar-shaped böreks filled with usually feta cheese and parsley. They are baked in a wood-fired oven until golden brown and crispy.

Su Böreği (Boiled Yufka Layers with Cheese and Parsley): This börek type made with boiled egg-based yufka sheets, layered with cheese and parsley, and then baked in a wood-fired oven. This results in a softer, moister texture compared to other böreks and highly popular as snacks and as breakfast.

Etli Patatesli Turta (Meat and Potato Pie): A savoury pie made with a shortcrust pastry filled with a mixture of meat (typically beef) and potatoes, sometimes with added vegetables. Baking this pie in a wood-fired oven adds a unique depth of flavour to the dish.

Desserts
Baklava Types:

Bülbül Yuvası (Nightingale's Nest): This type of baklava is made by rolling phyllo dough around a filling of crushed pistachios or walnuts, creating a cylindrical or bird's nest shape. It is then baked and soaked in sugar syrup.

Cevizli Baklava (Walnut Baklava): Made with layers of phyllo pastry filled with crushed walnuts, it is one of the most popular types of baklava in Türkiye.

Çikolatalı Baklava (Chocolate Baklava): A modern twist on traditional baklava, which incorporates chocolate into the layers of phyllo dough or as a drizzle on top of the dessert.

Çıtır Baklava (Crispy Baklava): A type of baklava made with especially thin phyllo dough layers, resulting in a lighter, crispier texture compared to other baklava types.

Dilber Dudağı Baklava (Lady's Lips or Sweetheart's Lips Baklava): Its name comes from the shape of the final pastry. The phyllo dough is filled with crushed nuts (usually pistachios or walnuts) and then folded into a half-moon or lip-like shape.

Fındıklı Baklava (Hazelnut Baklava): This variation uses crushed hazelnuts as the main filling, offering a unique taste and texture compared to the more common pistachio and walnut versions.

Fıstık Ezmesi Baklava (Pistachio Paste Baklava): Instead of using crushed pistachios, this type of baklava uses a smooth pistachio paste as the filling, creating a distinct flavour and texture.

Fıstıklı Baklava (Pistachio Baklava): This type of baklava is made with layers of phyllo pastry filled with finely crushed pistachios and is one of the most popular types of baklava in Türkiye.

Gül Baklava (Rose Baklava): This visually appealing baklava is shaped like a rose and often filled with pistachios or walnuts.

Havuç Dilimi (Carrot Slice): This baklava is shaped like a triangle or a carrot slice, hence the name. It is usually filled with pistachios.

Baklava Types:

Kaymak Baklavası (Clotted Cream Baklava): Layers of phyllo dough are filled with a combination of crushed nuts (pistachios, walnuts, or hazelnuts) and kaymak, which gives the dessert a creamy texture.

TABLE 6.12 *(Continued)* Pastries and Desserts in Turkish Cuisines that can be cooked in wood-fired oven

Kuru Baklava (Dry Baklava): Unlike most baklava types, kuru baklava is not soaked in syrup. Instead, it is baked with a minimal amount of sugar or syrup, giving it a crumbly, less sweet texture.

Saray Sarması (Palace Wrap): This baklava is made by rolling phyllo dough around a filling of crushed nuts, creating a spiral shape. It is then baked and soaked in sugar syrup.

Şöbiyet: A type of baklava made with layers of thin, buttery phyllo dough filled with a mixture of crushed pistachios or walnuts. It is baked in a wood-fired oven until golden and crispy. There is also a thick cream or kaymak version of the dessert, which gives it a richer taste and creamier texture.

Sütlü Nuriye (Milky Nuriye): This type of baklava incorporates milk into the sugar syrup, which results in a lighter, less sweet dessert. In the original version of the dessert, thinly sliced hazelnuts are used as fillings.

Kadayıf Types:

Borma (or Burma): This Middle Eastern dessert shares similarities with baklava. Instead of being flat and layered, borma uses kadayıf that is wrapped around a nut filling, typically pistachios or walnuts. The rolled pastries are then arranged in a round or rectangular shape, baked until golden, and soaked in a sweet syrup, often flavored with rose water or orange blossom water.

Burma Kadayıf (Rolled Kadayıf): This dessert is made by rolling the shredded kadayıf around a filling of finely crushed pistachios or walnuts, forming a spiral shape. The rolls are baked and then soaked in a sugar syrup.

Cennet Camuru (Heaven's Mud): The kadayıf and nuts (crushed walnuts or pistachios) are mixed together and pressed into a baking dish before being baked.

Ekmek Kadayıfı (Bread Kadayıf): A bread pudding made from kadayıf dough that is layered with a mixture of nuts and baked in the oven. It is served with kaymak (clotted cream) or ice cream.

Kadayıf Dolması (Rolled or Kadayıf, Kunefe variation): This dessert involves wrapping the shredded dough around a sweet or savory filling, such as sweetened nuts or cheese, and then baking until crispy. The finished rolls are soaked in a sugar syrup.

Künefe: A sweet, cheese-filled dessert made by sandwiching a layer of unsalted cheese between two layers of shredded kadayıf. The dessert is then baked or fried until crispy and golden, and soaked in a sugar syrup flavoured with lemon juice or rose water. It is served generously topped with crushed pistachios.

Tel Kadayıf (Thin Kadayıf): In this well-known typical kadayıf dessert, kadayıf is mixed thoroughly with melted butter. Then the crushed walnuts are spread between two layers of kadayıf and compacted before being baked.

Puddings:

Asure (Noah's Pudding): It is made with a mixture of grains, legumes, dried fruits, and nuts. It is sweetened with sugar and flavoured with spices like cinnamon and cloves.

Fırında Sütlaç (Oven-Baked Rice Pudding): Sütlaç is a traditional Turkish rice pudding made from rice, milk, sugar, and flavoured with vanilla. It is baked in the oven to create a golden-brown crust on top and served with a dusting of cinnamon. The dessert is typically served in individual clay pots or ramekins.

Puddings:

Kazandibi (Bottom-Caramelized Milk Pudding): A caramelized milk pudding characterized by its dark, caramelized bottom layer. It is made with milk, sugar, and thickening agents like rice flour or corn starch and often dusted with cinnamon before serving.

Keskul (Almond Milk Pudding): This milk pudding is made with crushed almonds, milk, sugar, and a thickening agent like corn starch or rice flour. It is garnished with coconut flakes, crushed pistachios, or cinnamon.

Muhallebi (Milk Pudding): A Middle Eastern-origin milk pudding highly popular in Türkiye. It is made with milk, sugar, and corn starch or rice flour and flavoured with rose water or orange blossom water. It is typically garnished with ground cinnamon or crushed nuts.

(Continued)

TABLE 6.12 (Continued)   Pastries and Desserts in Turkish Cuisines that can be cooked in wood-fired oven

Tavuk Göğsü (Chicken Breast Pudding): A unique Turkish pudding made from shredded chicken breast, milk, sugar, and a thickening agent like rice flour or corn starch. The chicken adds texture but has a neutral taste, making it a surprising yet delicious dessert.

Fruit Desserts:

Armut Tatlısı (Pear Dessert): Peeled pears are simmered in a mixture of sugar, water, and spices like cinnamon, cardamom, and star anise until tender. The poaching liquid is then reduced to a syrup and poured over the pears before serving.

Ayva Tatlısı (Quince Dessert): Quinces are peeled, halved, and cored, then cooked with sugar, water, and cloves or cinnamon. The hollowed-out core is sometimes filled with a mixture of sugar, walnuts, or clotted cream. The dessert is served with syrup and often accompanied by a dollop of kaymak or ice cream.

Elma Tatlısı (Apple Dessert): The apples are cored and filled with a mixture of nuts, raisins, cinnamon, cloves, and nutmeg. They are then baked until tender and often served with syrup, honey, or yogurt.

İncir Tatlısı (Fig Dessert): Dried figs are filled with walnuts or pistachios, simmered in a syrup flavoured with lemon juice and cinnamon, and served with clotted cream or yogurt.

Kabak Tatlısı (Pumpkin Dessert): Pumpkin is sliced (about 3cm thick) and cubed (to one or two bite-sized pieces) and simmered in granulated sugar until it releases its juice (usually overnight). Then it is gently cooked in a wood-fired oven until tender. It can be served warm or cold after being garnished with crushed walnuts or tahini and served with kaymak or clotted cream.

For the selection of pumpkin, a firm, sweet flesh and a smooth texture pumpkin is ideal. However, many types of pumpkins are suitable for the dessert, which comes in different regional names. These include Sugar Pumpkin (Pie Pumpkin), Kabocha, Butternut Squash, Cinderella pumpkin, Blue Hokkaido Pumpkin, Butternut Pumpkin, Turk's Turban Pumpkin, Queensland Blue Pumpkin, Jap Pumpkin (Kent Pumpkin), Golden Nugget Pumpkin, and Dumpling Pumpkin.

Kayısı Tatlısı (Apricot Dessert): This popular Turkish dessert is made with dried or fresh apricots that are boiled in sugar syrup and filled with clotted cream or crushed nuts, then served chilled.

Kestane Tatlısı/Kestane Şekeri (Chestnut Dessert/Chestnut Candy): Chestnut desserts are popular in Turkish cuisine, especially during the fall and winter months. In the preparation of the dessert, chestnuts are boiled in sugar syrup until tender and then allowed to cool and crystallize.

Combined Desserts:

Acıbadem Kurabiyesi (Almond Cookies): Made with ground almonds, sugar, and egg whites, these cookies are a popular treat in Turkish cuisine. They can be baked in a wood-fired oven for a crispy, smoky, and nutty flavour.

Fırında Sıcak Helva (Baked Warm Halva): The main ingredients include semolina, unsalted butter, granulated sugar, water, milk, and crushed walnuts or hazelnuts, with ground cinnamon for garnish. A second version of this dessert, named Sesame Seed Halva (or Tahini Halva), replaces semolina, sugar, water, and milk with pre-made tahini halva as the base. This halva type also mixes lemon zest and lemon juice into the melted butter.

Peynirli Helva (Cheesy Halva): A sweet cheese-based dessert made with unsalted fresh cheese, sugar, semolina, and butter. It can be baked in a wood-fired oven to create a slightly caramelized crust on top while maintaining a creamy, smooth texture inside.

Pide Tatlisi (Pide-Based Dessert): The popular Turkish flatbread, Pide, can be filled with sweet ingredients like cheese, fruits, chocolate, nuts, and tahini, then baked in a wood-fired oven. The results are a delicious, crispy, and flavourful treat.

Poğaça (Filled Bread Roll): Although poğaça is usually a savoury pastry, it can be made with sweet fillings like fruit preserves, sweetened cheese, or chocolate. The dough is often leavened with yeast and is thicker than phyllo or yufka, resulting in a softer, bread-like texture.

TABLE 6.12 (Continued) Pastries and Desserts in Turkish Cuisines that can be cooked in wood-fired oven

Vezir Parmagi (Vizier's Finger): It is made from a dough of semolina, flour, yoghurt, and eggs. The dough is shaped into small cylinders, and a sweet filling, typically made from crushed walnuts or pistachios mixed with sugar and cinnamon, is placed in the center. The filled pastries can be cooked in the oven until golden brown and soaked in a sugar syrup flavoured with lemon juice or rose water to give them a sweet and sticky texture.

A List of Other Turkish Desserts Suitable for Wood-Fired Oven Cooking

Bademli Sam Tatlisi (Almond Semolina Dessert)	Portakalli Revani (Orange Semolina Syrup Cake)
Peynirli Kek (Cheesecake)	Revani (Semolina Syrup Cake)
Taze ve Kuru Incir Tatlisi (Fresh and Dry Fig Dessert)	Sekerpare (Sweet Cookie Bites or Sweet Pastry Bites)
Hira Tatlisi (Hira's Semolina Dessert)	Şıllık (Şıllıki) Tatlısı (Şıllık Dessert)
Hurmisa (Cerkes Tatlisi)	Sultan Tatlısı (Sultan's Dessert)
Irmik Tatlisi (Semolina Dessert)	Sut Helvasi (Milk Halva)
Irmikli Sekerpare (Sweet Morsels or Sweet Bites with Semolina)	Sutlu Kadayif (Milky Kadayif)
Laz Boregi (Laz Pastry)	Sutlu Revani (Milky
Muhallebili Kadayif (Kadayif with Milk Pudding)	Taş Ekmeği (Stone Bread Dessert)
	Telli Baba Tatlisi (Telli Baba Dessert)
	Yogurt Tatlisi (Yoghurt Dessert)

6.8.2 Baklava

It is worth noting that despite the diverse range of desserts and their distinct shapes and textures, they often share similar ingredients limited to yufka (phyllo pastry), kadayif (shredded phyllo dough), sugar or honey, milk, cheese, nuts, coconut powder, rice, and rice powder.

Baklava is the most well-known and popular dessert originating from the Ottoman Empire, influenced by Middle Eastern and Mediterranean cuisines. Although there are various types of baklava, they all consist of layers of thin, flaky phyllo pastry filled with a mixture of finely crushed or sliced nuts (such as walnuts, pistachios, hazelnuts, peanuts, and almonds) and kaymak (clotted cream).

In the classic flat-style baklava, before baking, the phyllo layers are multi-layered and cut diagonally in parallel lines, resulting in a bite-sized diamond or rhomboid shapes, squares, or rectangles. These patterns create an appealing presentation and individual portions. The purpose is to allow melted purified butter (ghee) to seep between the layers before baking and to enable the layers to soak in syrup after baking. The syrup, made from sugar or honey, is sometimes flavoured with cinnamon, cloves, lemon, or rose water. It is important to ensure that the syrup is cooler while the baklava is warm, or vice versa, during the preparation process.

It is noteworthy that there are many other desserts worldwide associated with or similar to baklava, characterized by flaky pastry layers. These include Saragli and Bougatsa (Greek), Paklava (Azerbaijan and Armenia), Sfogliatelle (Italian), and Borma (known as Burma in Turkey) in Lebanon and Syria.

6.8.3 Kadayif (or Kadaif)

Kadayif is a popular ingredient in Turkish and Middle Eastern desserts. It is made from "kadayif," which can be prepared using two different methods:

- Thinly rolling out phyllo dough to a yufka shape and cutting it into fine strands.
- Using a special spinning machine with a rotating drum that has small holes and a heated surface to form and cook the kadayif.
- These strands of dough are used to create a crispy, golden-brown crust in various Turkish dessert combinations.

6.8.4 Puddings

In Turkish cuisine, "pudding" typically refers to a sweet dessert with a smooth, creamy, or semi-solid texture. These puddings are primarily milk-based and thickened with rice flour, corn starch, or wheat starch. They are sweetened with sugar and flavoured with cinnamon, vanilla or cloves. Turkish puddings are usually cooked in individual small cups or clay pots, or in shallow large trays that are later divided into squares. They can be served cold or warm and are often garnished with nuts or fruits. Ingredients such as rose water or orange blossom water may be incorporated for a distinct aroma.

6.8.5 Fruit Desserts

Fruit desserts are highly popular in Turkish and Middle Eastern cooking. They involve the use of fresh, dried, or preserved fruits combined with various spices and sweeteners. Cooking these desserts in a wood-fired oven enhances their unique flavours and textures. While the choice of fruit may vary, these desserts often share elements such as syrups, spices, nuts, or dairy products to enhance flavour and texture. One example is Elmali Turta, also known as Turkish Apple Pie, which features a buttery pastry crust and a filling made from apples, sugar, and cinnamon.

When cooking desserts and pastries in a wood-fired oven, it is crucial to monitor the temperature closely to ensure even baking without burning. Desserts require lower temperatures than savoury dishes. It is recommended to cook desserts when the active fire is out, and the oven is in the cooling mode. The Data Logging Chapter provides insights on achieving the necessary temperature range for cooking pastries and desserts. Please refer to Table 6.12 for a list of Turkish pastries and desserts that can be cooked in a wood-fired oven.

6.9 MARINATION

In many cuisines, marinades are commonly used to tenderize meat, making tough cuts more palatable, and/or to enhance flavours, which can significantly reduce cooking time. However, in slow cooking with a wood-fired oven, marination is not required if the goal is solely to tenderize the meat.

In Turkish cooking, marinades often include acidic substances like green grape or green tomato juice, vinegar, citrus juice, pickle juice, or buttermilk. Additional components may consist of salt, chilli powder, peppercorns, ginger, garlic, curry paste, tamarind paste, mustard, and a variety of fresh or dried herbs. Sweeteners like ketchup, honey, agave, barbecue sauce, pomegranate molasses, and soft drinks, as well as salty substances such as pickle juice and sea salt, and fatty substances like olive oil, sesame oil, yoghurt, milk, tahini, and mayonnaise, are also used.

In Turkish cuisine, sugar is rarely used for meat marination, but beer and wine have been introduced in recent years. Exotic substances or enzymes like mango, papaya, kiwi fruit, miso, soy sauce, or fish sauce, which are common in Asian cuisines as salty substances, are not used in Turkish cooking. However, they may be incorporated in fusion cooking.

A typical marinade mix for meat dishes, such as shish kebabs, can include the following:

- 2 medium-sized onions, grated, and juice collected by pressing
- 4 large garlic cloves, crushed
- 1 tbsp black pepper
- 1 tsp finely crushed dried oregano
- 1/4 cup of olive oil
- Sufficient amount of milk to cover the meat being marinated.

- Combine all the ingredients in a container and soak the meat in this seasoned liquid. Cover the container and refrigerate overnight or for about 12 hours to prevent bacterial growth. The meat is then ready for cooking.

6.10 HERBS AND SPICES

As well known, both herbs and spices are plant-based ingredients. Herbs primarily refer to the leaves of plants and are commonly used fresh or dried, imparting a mild flavour in Turkish, Mediterranean, and European cuisines. On the other hand, spices are derived from various parts of plants, such as seeds, roots, bark, or fruits. They are typically used in powdered or crushed form, offering a strong and pungent flavour that adds warmth, depth, and complexity to dishes. Spices are commonly found in Turkish, Middle Eastern, Indian, and Asian cuisines.

Throughout the development of civilizations, ancient Egyptians, Greeks, and Romans embraced the use of spices and herbs in their cuisine, medicine, and rituals. These valuable commodities were traded through various regions, facilitated by trade routes like the Silk Road that connected Asia to Europe through Türkiye. This exchange allowed for the introduction of major spices like cinnamon, cassia, and black pepper. The abundance of spices and herbs found in present-day Türkiye is a testament to its rich culinary heritage and the influence of diverse cultures throughout history.

In Turkish cuisine, spices play a crucial role in creating flavourful dishes and sauces with unique blends of flavours and aromas. The significance of spices and herbs is evident in the vibrant marketplace known as the Egyptian Bazaar (Spice Bazaar or Mısır Çarşısı in Turkish) in Istanbul (Figure 6.9). Constructed in the 17th century, this bazaar originally served as a hub for the sale of spices and other goods.

FIGURE 6.9   A view of spices from Egyptian bazaar in Istanbul/Türkiye.

Ertuğrul, Nesimi, Personal photograph, Adelaide, Australia, 2022.

Table 6.13 provides a summary of the spices and herbs that enrich the diverse flavours found in Turkish cuisine. The table also offers a brief description of each spice and herb, highlighting their distinct characteristics and common uses.

TABLE 6.13 Commonly used spices and herbs in Turkish cuisine with Turkish names and their characteristic features and the recipes used

SPICE/HERB	DESCRIPTION	USE IN RECIPES
Aleppo Pepper or Red Pepper Flakes (Pul Biber)	A mild chilli pepper used to add heat and colour	Soups, stews, meat dishes
Allspice (Yenibahar)	Combined flavour of cinnamon, cloves, and nutmeg	Meat dishes, stews, soups, desserts
Anise (Anason)	Licorice-flavoured spice	Desserts, bread, alcoholic drinks
Basil (Fesleğen)	Popular herb	Tomato-based sauces, salads, garnish
Bay Leaves, Laurel (Defne yaprağı)	Aromatic herb	Soups, stews, meat dishes
Bean Herb (Fasulye Out)	Leaves or tendrils of various bean plants	Salads, garnish
Black Cumin (Kara kimyon)	Pungent, earthy spice	Bread, cheese, meat dishes
Black Pepper (Karabiber)	Versatile and widely used spice	Various dishes
Borage (Hodan otu)	Cucumber-like herb	Salads, teas, garnish
Capers (Kapari)	Pickled condiment or ingredient	Sauces, salads, fish dishes
Caraway Seeds (Çörek otu)	Warm, slightly sweet spice	Bread, desserts, meat dishes
Cardamom (Kakule)	Warm, aromatic spice	Desserts, rice dishes, beverages
Celery Leaf (Kereviz yaprağı)	Leaves of the celery plant	Salads, soups, stews
Chervil (Frenk maydanozu)	Delicate, parsley-like herb	Salads, soups, sauces
Chilli Pepper (Acı Biber)	Pungent flavour and spiciness	Various dishes
Chives, Garlic Chives (Sarımsak Otu)	Mild onion-garlic flavour	Salads, soups, omelettes
Cinnamon (Tarçın)	Sweet and warm spice	Savoury dishes, desserts
Cloves (Karanfil)	Strong, sweet spice	Rice dishes, desserts, beverages
Coriander (Kişniş)	Citrusy, slightly sweet spice	Spice blends, meat dishes, salads
Cumin (Kimyon)	Earthy, earthy spice	Meat dishes, soups, stews, rice dishes
Curry Powder (Köri Tozu)	Blend with ingredients such as turmeric, coriander, cumin, fenugreek, and chilli powder	Curries, soups, stews, rice
Dill Seeds (Dereotu Tohumu)	Slightly bitter, pungent, and warm flavour	Pickling recipes, bread, spice blends, soups, stews, sauces
Dill (Dereotu)	Fragrant, slightly bitter herb	Salads, soups, sauces, fish dishes, yoghurt-based dishes
Fennel Seeds (Rezene)	Licorice-flavoured spice	Desserts, bread, fish dishes
Fenugreek (Çemen otu)	Slightly bitter spice	Pickles, bread, spice blends
Galangal (Havlıcan)	Ginger-like spice	Soups, stews, meat dishes

TABLE 6.13 (Continued)   Commonly used spices and herbs in Turkish cuisine with Turkish names and their characteristic features and the recipes used

SPICE/HERB	DESCRIPTION	USE IN RECIPES
Ginger (Zencefil)	Warm, pungent, and slightly sweet flavour	Stir-fries, curries, soups, baked goods, beverages
Juniper Berries (Ardıç Tohumu)	Pine-like flavour	Meat dishes, stews, marinades
Lemon Thyme (Limon Kekiği)	Lemon-scented variety of thyme	Fish dishes, salads, teas
Lemon Balm (Melisa Otu, Oğul otu)	Lemon-scented herb	Teas, beverages, desserts, salads,
Lemon Basil (Limon Fesleğen)	Citrusy herb	Salads, pasta, seafood, chicken, infused oils, dressings, marinades
Lemon Mint (Limon Nanesi)	Minty, lemony aroma and flavour	Teas, cocktails, desserts, garnish
Lemon Verbena (Limonsuktu)	Lemon-scented herb	Flavouring agent in dishes, desserts, beverages
Lovage Leaves, Maggie Plants (Yaban Kerevizi)	Celery-flavoured herb	Soups, stews, meat dishes, salads
Mahaleb (Mahlep)	Sweet, slightly sour spice from St. Lucie cherry tree	Desserts, baked goods
Marjoram (Mercanköşk)	Fragrant herb similar to oregano	Meat dishes, stews, soups
Mastic (Damla sakızı)	Piney, resinous spice	Desserts, ice cream, beverages
Mint (Nane)	Refreshing herb	Salads, yoghurt-based dishes, beverages
Mint-scented Geranium (Fesleğenli Pelargon)	Minty, aromatic herb	Flavouring teas, desserts, beverages, garnish
Mustard Seeds (Hardal tohumu)	Spicy seeds	Pickles, condiment in meat dishes
Nigella Seeds (Çörek otu)	Small, black seeds with onion-like flavour	Bread and pastries
Nutmeg (Muskat)	Warm and sweet spice	Desserts and beverages
Oregano (Kekik)	Popular herb	Meat dishes, salads, stews, tomato-based sauces
Paprika (Toz Biber)	Sweet or hot spice from ground peppers	Soups, stews, meat dishes
Parsley (Maydanoz)	Popular herb	Garnish, salads, soups, meat dishes
Poppy Seeds (Haşhaş Tohumu)	Small, nutty seeds	Bread, pastries, desserts
Purslane (Semizotu)	Slightly sour, succulent herb	Salads, soups, stews
Rosemary (Biberiye)	Fragrant, pine-flavoured herb	Meat dishes, stews, soups
Ras El Hanout (Ras El Hanout)	Spice blend used in North African cuisine	Stews, tagines, couscous, rice
Rocket/Arugula (Roka)	Peppery, slightly bitter herb	Salads, garnish, pasta, soups, sauces
Rose Petals (Gül Yaprağı)	Sweet, floral flavour	Desserts, beverages, jams, spice blends

(Continued)

TABLE 6.13 (Continued)   Commonly used spices and herbs in Turkish cuisine with Turkish names and their characteristic features and the recipes used

SPICE/HERB	DESCRIPTION	USE IN RECIPES
Saffron (Safran)	Highly prized and expensive spice	Rice dishes, desserts, beverages
Sage (Adaçayı)	Strong, slightly bitter herb	Meat dishes, stews, teas
Sahlep (Salep)	Flour from wild orchid tubers	Sahlep, a hot beverage; thickening agent in Turkish ice cream
Saltbush Dukkah (Tuzlu Çalı Dukka)	Egyptian spice blend	Seasoning and garnish
Savory (Biberiye)	Peppery, slightly bitter herb	Meat dishes, stews, soups
Sesame Seeds (Susam)	Nutty seeds	Garnish, tahini, toppings in bread products
Smoked Paprika (Közlenmiş Biber Tozu)	Ground, smoked peppers	Stews, soups, sauces, meat rubs
Sorrel (Kuzukulağı)	Tangy, slightly sour herb	Salads, soups, sauces
Spearmint (Nane)	Fresh, slightly sweet herb	Salads, beverages, garnish
Star Anise (Badyan/Yıldız Anason)	Sweet, licorice-like flavour	Chinese, Vietnamese, Indian cooking
Sumac (Sumak)	Tangy, lemony spice	Salads, meat dishes, garnish
Sweet Marjoram (Mercanköşk)	Sweet, mild, slightly citrusy flavour with a hint of pine	Soups, stews, sauces, meat preparations
Szechuan Pepper (Szechuan Biberi)	Mildly spicy, woody, citrusy taste with numbing sensation	Stir-fries, soups, spice blends
Tarragon (Tarhun)	Sweet, slightly licorice-flavoured herb	Fish dishes, salads, sauces
Thyme (Zahter or Kekik)	Fragrant herb	Meat dishes, stews, soups
Turmeric (Zerdeçal)	Warm, earthy, slightly bitter spice	Various cuisines and dishes
Urfa Pepper (Isot)	Smoky, raisin-like, mildly spicy Turkish chilli pepper	Kebabs, stews, salads, baharat spice blend
Vanilla (Vanilya)	Sweet, aromatic flavouring	Desserts, baked goods, beverages
Watercress (Turp Otu)	Peppery, slightly bitter herb	Salads, soups, sandwiches, garnish
Wild Garlic (Yabani Sarımsak)	Milder, delicate garlic flavour	Salads, pestos, soups, sauces
Wild Thyme (Dağ Kekigi)	Strong, aromatic herb with grey-green leaves	Stews, soups, meat preparations
Za'atar (Zahter)	Earthy, tangy, slightly nutty spice blend	Seasoning, dipping sauce, meat, vegetables, bread

The main purpose of the above table is to highlight the traditional pairings of dishes with specific herbs and spices, allowing for the identification of groups of dishes that share similar flavour profiles. In a short summary, black pepper and cinnamon are widely used in various Turkish dishes, while cumin is a staple spice in meat products. It is worth noting that ginger is less commonly used in Turkish cuisine. To further illustrate these points, Figures 6.10 and 6.11 showcase the commonly used spices, as well as green and dry herbs, in Turkish cuisines.

FIGURE 6.10   Commonly used spices in Turkish cuisine from left to right: black pepper (kara biber), cardamom (kakule), cinnamon sticks (tarçın), cloves (karanfil), safflower (aspir), saffron (safran), coriander seeds (kişniş), cumin seeds (kimyon), fennel seeds (rezene), mustard seeds brown (hardal tohumu), mustard seeds yellow (hardal tohumu), a blend of powdered thyme (zahter), nigella seeds (çörek otu), nutmeg whole (muskat), fennel seeds (rezene), paprika (toz biber), saltbrush dukka (tuzlu çalı dukka), and aniseed (anason).

Ertuğrul, Nesimi, Personal photograph, Adelaide, Australia, 2022.

FIGURE 6.11   Commonly used green and dry herbs in Turkish cuisine from left to right: sweet basil (feslegen), thyme (kekik), mint (nane), sweet marjoram (mercankosk), tarragon (tarhun), parsley (maydanoz), daphne (defne), ginger (zencefil), dry mint leaves (nane), dry curry leaves (kori), dry oregano leaves (kekik), and dry marjorm leaves (mercanköşk).

Ertuğrul, Nesimi, Personal photograph, Adelaide, Australia, 2022.

REFERENCES

[1] Ethnic Vahid Mohammadpour Karizaki and Traditional Iranian Breads. Different types, and historical and cultural aspects. *Journal of Ethnic Foods* 4(1), 8–14 (2017), Issn 2352-6181. 10.1016/J.Jef.2017.01.002

[2] Türk Mutfağina Özgü Dolma ve Sarmalar and T. C. Millî Eğitim Bakanliği Yiyecek İçecek Hizmetleri, Ankara (2018), available at http://www.Megep.Meb.Gov.Tr/Mte_Program_Modul/Moduller_Pdf/T%C3% Bcrk%20mutfa%C4%9f%C4%B1na%20%C3%96zg%C3%Bc%20dolma%20ve%20sarmalar.Pdf, accessed on Sept 20/2023.

[3] Ekmek Çeşitleri and T. C. Millî Eğitim Bakanliği Yiyecek İçecek Hizmetleri, Ankara (2018), available at http://www.megep.meb.gov.tr/mte_program_modul/moduller_pdf/Ekmek%20%C3%87e%C5%9Fitleri. pdf, accessed on 10/05/2023.

[4] Fodla Bread, Mavi Boncuk Cornucopia of Ottomania and Turcomania (2023), available at https:// Maviboncuk.Blogspot.Com/2022/02/Fodla-Bread.Html, accessed on 10/05/2023.

[5] Halıcı, F. Geleneksel Türk yemekleri ve beslenme: Geleneksel Türk Mutfağı Sempozyumu bildirileri; 10 - 11 Eylül 1982 Konya Turizm Derneği Yayınları, by Feyzi Halıcı and Konya Kültür ve Turizm Derneği, 270 pages .1982.

[6] Mahmut Tezcan, Grains and Breads and Other Dough-Based Foods in Turkish Culinary Culture (2023), available at http://www.turkish-cuisine.org/ingredients-7/ingredients-used-in-turkish-cuisine-66/grains-and-breads-234.html?PagingIndex=4, accessed on 10/05/2023.

[7] Anadolu Mutfak Kültüründen Esintiler, Konya Büyükşehir Belediyesi Kültür Yayınları: 419, Konya, Turkiye (2020), ISBN: 978-605-389-365-3, available at file:///C:/Users/a1002260/Downloads/Turk_ Mutfak_Kulturunde_Ekmek.pdf, accessed on 10/05/2023.

[8] Yıkmış, S., Sağlam, K. and Yetim, A. The examination of spices used in the Ottoman palace cuisine. *Journal of Human Sciences* (March 2017). 10.14687/Jhs.V14i1.4508, available at https://Www.Researchgate. Net/Publication/315917500, accessed on 10/05/2023.

[9] Ayyıldız, S. and Sarper, F. Antioksidan baharatların osmanlı saray mutfağındaki yeri. *Journal of Humanities and Tourism Research* 9(1), 363–380 (March 2019). 10.14230/joiss665

[10] Demirgül, F. Çadırdan Saraya Türk Mutfağı. *International Journal of Turkic World Tourism Studies* 3(1), 105–125, (July 2018).

[11] Saçikarali, M. *Türk Mutfak Kültürümüzde Aşurenin Tarihsel Süreci, Yüksek Lisans Tezi* (Master Thesis). Aralik: Gaziantep Üniversitesi, Gaziantep/Turkiye, 2015.

[12] Al-Faqih K. *170 Fresh and Healthy Mediterranean Favorites* (1st Ed.). Three Forks, September 1, 2009. ISBN-10: 0762752785 and ISBN-13: 978-0762752782.

Selected Recipes for the Wood-Fired Ovens

7

To explore the richness of Turkish and the World cuisines further in Wood-Fired Ovens, please visit: http://woodfirewonders.com/book-recipies/

7.1 INTRODUCTION

Food has always played an integral role in human culture, connecting individuals and communities through shared meals and cherished traditions. Each cuisine tells a captivating story, reflecting the unique blend of history, geography, climate, and customs that shape its region.

Therefore, in this chapter, a selection of sample recipes from the cuisines covered in Chapter 6 are included. These recipes have been structured and defined to take full advantage of the distinctive characteristics of wood-fired oven cooking, including optimal temperature ranges and suitable locations within the oven. Drawing from the valuable insights shared in the Data Logging Chapter, these recipes are tailored to deliver flavoursome results.

While the recipes featured in this chapter encompass a wide range of flavours, techniques, and ingredients, offering a hint into the culinary treasures of Turkish cuisines and fusion cooking, they are designed with the expectation that the wood-fired oven will infuse each dish with a distinct smoky flavour and enhance their textures, elevating the overall cooking experience.

Furthermore, before starting the recipes, a comprehensive guide to preparing associated dough products is covered. Therefore, in-depth discussions on leavening agents, techniques for creating wild yeast, the art of wheat-based dough making, ingredient ratios, various dough mixing methods, and their impacts on the final product are all covered. Moreover, techniques for testing dough readiness are explored and measurement units and conversions to ensure precision and consistency in recipe preparations are included.

To further enrich the culinary journey on the wood-fired oven cooking, use the Web Link and a QR code given to access a comprehensive collection of recipes, primarily focused on Turkish cuisines while featuring a range of dishes from around the world. All the recipes are specifically selected and designed for wood-fired oven cooking. The online resource will be regularly updated, continuously expanding the culinary repertoire while contributing to the art of cooking with a wood-fired oven.

DOI: 10.1201/9781032640136-7

7.2 GETTING READY FOR DOUGH PRODUCTS

7.2.1 Leavening

As previously mentioned, Turkish bread products can be either leavened or unleavened. Leavened breads and baked goods use leavening agents for rising and textural development, falling into three groups [1]:

- Biological leavening: living microorganisms like yeast or bacteria produce carbon dioxide, causing dough expansion and rising. Common examples include sourdough bread (fermented flour-water mixture with wild yeast and lactic acid bacteria) and yeast bread (using commercial yeast, Saccharomyces cerevisiae).
- Chemical leavening: acid–base reactions generate carbon dioxide, leading to dough rising. Typical agents are baking soda (base) and baking powder (base–acid mix). Examples include quick breads (banana bread, cornbread, muffins), cakes (layer cakes, cupcakes), and cookies (chocolate chip, sugar cookies).
- Mechanical leavening: physically incorporating air or steam into dough/batter, which expands during baking, causing rising. Examples are puff pastry (dough folded and rolled multiple times, with thin layers separated by butter), choux pastry (flour-water cooked dough with added eggs), and whisked sponge cakes (air incorporation from whisking eggs and sugar).

In addition, baking often employs a blend of leavening methods, and accommodates five common types of yeast including [2]:

- Active Dry Yeast: dehydrated yeast that requires rehydration and activation before use, known for its long shelf life and consistent performance.
- Instant Yeast: also called rapid-rise or fast-acting yeast, a fine-grained dry yeast that doesn't need proofing and is suitable for recipes with shorter rising times.
- Fresh Yeast: a moist, perishable yeast form used by professional bakers for its reliable results; however, its short shelf life and temperature sensitivity make it less popular for home baking.
- Wild Yeast: naturally occurring yeast found in the environment, captured and cultivated through a sourdough starter for fermentation and flavour development in sourdough bread and other baked goods.
- Nutritional Yeast: a deactivated yeast used as a food supplement or seasoning, offering a savoury, cheesy flavour and nutritional benefits, but not typically used in baking.

7.2.1.1 Making Wild Yeast, Wheat-Based

Wild yeast is a naturally occurring yeast found in the environment. Wild yeast can be captured and cultivated by creating a sourdough starter, which is a mixture of flour and water that ferments over time. To create your own wild yeast sourdough starter, you will require one cup (120g) all-purpose or whole wheat flour, one cup (240ml) water (preferably filtered or distilled), and a glass or plastic container (avoid using metal) with a loose-fitting lid or cloth and follow the below steps.

- Combine equal parts flour and water in a container to form a smooth, thick batter, and cover the container with a loose lid or a cloth to let it breathe. Then place the container in a warm spot with a temperature of 20–25°C for optimal yeast growth.
- After 24 hours, check for bubbles or activity. If visible, discard half and add fresh flour and water to the remaining mixture. If not, wait another day and check again.

- Continue daily feedings, discarding half the mixture and adding fresh flour and water each time.
- Over 5–7 days, the starter should become bubbly, double in size within 4–6 hours of feeding, and develop a tangy smell.
- Use the mature starter for baking and maintain it regularly. Store unused portions in the fridge, feeding them weekly. Allow the refrigerated starter to reach room temperature and feed it once before using it for baking.

7.2.1.2 Making Chickpea-Based Yeast

To make a chickpea-based sourdough starter using aquafaba (which is the viscous water in which chickpeas have been cooked), you'll need one cup (120g) of all-purpose or whole wheat flour, one cup (240ml) of aquafaba, and a non-metal container with a loose lid or clean cloth. Then follow the below steps:

- Mix flour and aquafaba in the container until smooth and thick, and cover the container with a loose lid or cloth to allow breathing.
- Place the container in a warm spot (21–24°C) to promote yeast growth.
- After 24 hours, if you see bubbles or activity, discard half the mixture and add 1/2 cup of flour and 1/2 cup of aquafaba. Stir well, and repeat daily feedings.
- In a few days, it should become bubbly with a sour aroma. It's mature when it consistently doubles in size within 4–6 hours of feeding and has a tangy smell.
- Continue regular feedings and store the starter in the fridge when not in use. Bring it to room temperature and feed once before using it for baking.

7.2.2 Dough Making

Dough making involves four main ingredients: flour, water, yeast, and salt, with proportions based on the flour's weight. The final baked product's weight is less than the initial total weight due to evaporation, and its texture depends on the size and number of air pockets formed, which are influenced by water content, yeast type, and fermentation duration.

Water temperature is essential for yeast activation and fermentation duration, with an ideal range of 40.5–43.3°C for doughs containing yeast. Warmer water may be used in cooler environments, while cooler water can be used in warmer environments. In warmer climates, ice cubes may be added during mixing to slow yeast activity.

For yeast-free doughs, water temperature impacts gluten development and consistency. Room temperature or slightly warmer water is recommended. Fermentation duration depends on dough temperature, with warmer dough fermenting faster. Typically, bread dough ferments for 1–2 hours at room temperature (around 24°C). If the dough's temperature differs from room temperature, adjust the fermentation time accordingly. Other factors, like yeast type and dough hydration, also affect fermentation time.

7.2.2.1 Ratios of Ingredients

The proportions of ingredients in dough-based recipes, often called Baker's percentage or Baker's math, are crucial for maintaining consistency, scaling, and comparing recipes. This standardized method expresses ingredient ratios relative to flour weight, with flour always at 100% and other ingredient proportions calculated accordingly.

Table 7.1 shows the proportions for four typical Turkish bread products. These examples are a starting point, and individual recipes may vary based on ingredients, techniques, or desired outcomes,

TABLE 7.1   The summary of the proportions of ingredients in four typical bread products

INGREDIENTS	FLATBREAD	PIDE	LOAF BREAD	FRESH YEAST BREAD
Flour	100%	100%	100%	100%
Water	60%	65%	70%	65%
Yeast (Dry)	1%	1.5%	2%	–
Yeast (Fresh)	–	–	–	3%
Salt	2%	2%	2%	2%
Sugar	–	2%	4%	–
Olive oil	3%	4%	–	–
Butter	–	–	5%	–

TABLE 7.2   Typical hydration percentages and unique characteristics of some dough products

PRODUCT	HYDRATION	CHARACTERISTICS
Bagel	50–60%	Dense and chewy texture
Pizza Dough	60–65%	Firm but pliable, easy to stretch, crisp yet tender crust
Sandwich Bread	60–70%	Soft and tender crumb, with a light and airy structure
Ciabatta	75–85%	Open crumb with large irregular holes, a crisp crust, and a slightly chewy texture
Focaccia	70–80%	Moist and tender with a slightly chewy texture, well-developed crumb structure, and a crisp crust
Sourdough Bread	65–80%	It can range from a moderately open crumb with a firmer crust to an open crumb with large irregular holes and a crisp, chewy crust.

especially in wood-fired ovens with structural variations and unpredictable heating characteristics. Note that if pide bread is used as a base, the final product's characteristics will also depend on the fillings.

Hydration is a key factor in dough preparation, as it influences the texture and properties of the baked goods. Expressed as a percentage, hydration impacts the dough's wetness or dryness, ultimately affecting the final product and cooking time. Table 7.2 provides examples of various dough products, their typical hydration percentages, and their unique characteristics.

Note that hydration percentages are general guidelines and may vary depending on factors such as ingredients, flour type, and desired results. Adjusting hydration levels can help achieve specific dough consistency, crumb structure, or crust texture.

7.2.2.2 Dough Mixing Methods and Their Impacts

The mixing method is vital for optimal dough quality, with factors like dough type, texture, and recipe requirements influencing the choice. Mixing methods, such as hand kneading, straight dough, and sponge-and-dough, impact dough development, gluten formation, and fermentation. Note that mixing affects the following parameters in dough making:

- Gluten development: insufficient mixing leads to weak gluten, causing dense, less voluminous bread, while overmixing breaks down gluten, resulting in a loss of elasticity and sticky or crumbly dough.
- Dough aeration: mixing incorporates air and distributes yeast evenly for consistent fermentation and rise. Inadequate mixing causes uneven aeration and inconsistent rise during baking.
- Dough temperature: mixing generates friction, raising dough temperature. Monitoring and controlling temperature during mixing is essential, as it influences yeast activity, fermentation rate, and dough consistency.

- Hydration: mixing affects water incorporation into the flour, impacting dough consistency and gluten development.
- Mixing time: undermixing results in insufficient gluten development and uneven ingredient distribution, while overmixing makes the dough overly elastic or sticky.

7.2.2.3 Testing Readiness of Final Dough

Assessing dough readiness is crucial for baking in wood-fired or conventional ovens. Two common methods for evaluating dough proofing are the "finger poke test" (earlobe test) and "windowpane test."

In the finger poke test, press a finger about 1.5 cm into the dough and observe its reaction. If it springs back immediately and completely, it needs more time to proof. If it slowly springs back halfway, it's ready to bake. If it doesn't spring back, it may be over-proofed, resulting in denser bread.

The finger poke test isn't as reliable as the windowpane test for gluten development but is helpful for assessing dough proofing.

In the windowpane test, perform the following steps on a small dough piece:

- Stretch the dough from the centre, forming a thin, translucent membrane.
- Hold it up to a light source.
- If it stretches evenly without tearing, the gluten network is well-developed, and the dough is ready for proofing and shaping. If it tears easily or doesn't form a thin membrane, more kneading is needed.

7.2.3 Measurement Units and Conversions

Wood-fired oven cooking is more rustic and relies on intuition, experience, and knowledge of the oven's built rather than precise measurements. This is due to the fluctuating temperatures and varied heat transfer methods in wood-fired ovens. However, it's still essential to follow recipes and measure ingredients that significantly impact taste and texture.

When cooking with a wood-fired oven, focus on ingredient proportions, cooking times, oven temperature, and heat transfer methods, while monitoring dishes closely. As you gain experience, you'll better understand how to adjust recipes for optimal results.

In this chapter, five key units for ingredient proportions and their conversions are provided in Table 7.3. Keep in mind that these conversions are approximate and can vary based on the substance and measuring tools. Despite being suitable for wood-fired cooking, the most accurate conversion should involve weight measurements.

TABLE 7.3 Some general equivalencies of cup, grams, teaspoons, tablespoons, and bunch of common ingredients

	CUP	GRAMS	TEA SPOON	TABLE SPOON	BUNCH
Flour	1	120–125	48	16	–
Water	1	237	48	16	–
Instant Yeast	–	9	3	1	–
Fresh Yeast	1	225	48	16	–
Olive Oil	1	215	48	16	–
Butter	1	230	48	16	
Sugar	1	200	48	16	–
Fresh Parsley	1	60	–	–	1

(Continued)

TABLE 7.3 (Continued)   Some general equivalencies of cup, grams, teaspoons, tablespoons, and bunch of common ingredients

	CUP	GRAMS	TEA SPOON	TABLE SPOON	BUNCH
Fresh Herbs	1	60	–	–	1
Dry Herbs	1	25–30	48	16	–
Fresh Spices	1	100–120	–	–	1
Dry Spices	1	120–150	48	16	–
Yogurt	1	250	48	16	–
Salt	1	290	48	16	–
1 Medium-sized onion	–	140	–	–	–
1 Medium-sized tomato	–	150	–	–	–
1 Medium-sized potato	–	150	–	–	–
1 Medium-sized garlic cloves	–	3	–	–	–
1 Medium-sized egg	–	50	–	–	–
Tomato paste	1	255	48	16	–

PS. 1 teaspoon = 5 ml; 1 tablespoon = 15 ml; 1 tablespoon= 3 teaspoons.

7.3 SAMPLE RECIPES

7.3.1 Plain Pide Bread (see Figure 7.1)

7.3.1.1 Ingredients

A 500g (4 cups) all-purpose flour plus extra for dusting, 7g (2 1/4 tsp) of instant yeast (or 9g, 2 1/4 tsp of active dry yeast), 10g (2 tsp) of sugar, 8g (1 1/2 tsp) of salt, 250ml (1 cup) warm water, 100g (1/3 cup) plain yogurt at room temperature, 60ml (1/4 cup) olive oil, 1 large egg, and white or black sesame or nigella seeds for topping.

7.3.1.2 Preparation and Cooking

- Combine flour, instant or active dry yeast, sugar, and salt in a large mixing bowl.
- Dissolve active dry yeast in warm water with sugar and let it sit for 5–10 minutes until frothy. Skip this step if using instant yeast.
- Make a well in the centre of the flour mixture, and add warm water (with dissolved yeast, if using), yogurt, and olive oil. Then gradually mix to form a sticky dough. Use a dough mixer for convenience.
- Knead the dough on a lightly floured surface for 10 minutes until smooth and elastic, and adjust the flour as needed.
- Place the dough in a lightly oiled bowl, cover, and let rest in a warm place for 1–2 hours until doubled in size.
- Deflate the dough, divide into 4–6 equal pieces, and roll out into 10–15mm thick ovals. Brush with egg wash and sprinkle with sesame seeds or other seeds.
- Bake at 270–290°C in a wood-fired oven for 8–10 minutes until golden brown and puffed.
- Serve warm with toppings or dips, or cool on a wire rack before storing.

FIGURE 7.1   Flat Plain Pide and two popular Turkish Loaves.

Ertuğrul, Nesimi, Personal photographtop.

7.3.1.3 Hints and Comments

- Make an egg wash with whisked egg, water, and/or yogurt. Alternatively, use a flour and water mixture for a low-cost option.
- A local (Divrigi/Türkiye) recipe of pide and bread dough includes as follows: 10 liters of water for 8kg of all-purpose flour, 250g of fresh yeast, and 100g of salt. The dough is mixed for 1/2 hour and then rested for 15 h (overnight). The same dough is used in two bread types: Pide bread is cooked at over 300°C and loaf bread is cooked at 120–250°C.

7.3.2 Turkish Loaf (see Figure 7.1 middle and bottom)

7.3.2.1 Ingredients

The 1 kg of all-purpose flour, 25g fresh yeast (or 15g dry yeast and 1 tsp of sugar), 15g of salt, 650 ml of warm water, and 2 tbsp of olive oil.

7.3.2.2 Preparation and Cooking

- Take the flour in a large mixing bowl or onto the bench top and create a flour well in the centre, then add water and olive oil and gently mix with the flour for 10 minutes or until the dough is smooth and elastic.
- Move the final dough onto the benchtop and knead it by slamming, folding, and stretching for about 10 minutes.
- Add salt and knead the dough for another 2 minutes until the salt is absorbed, and then add the crumbled yeast and knead it until incorporated into the dough.
- Place the dough in a lightly oiled bowl, cover with a kitchen towel or plastic wrap, and let it rest in a warm place for about 1 hour, or until it has doubled in size.
- Move the risen dough onto the benchtop and divide it into four (for the given ingredients) equal pieces, fold the edges inward to form balls, cover them with a kitchen towel, and allow it to rest for 20 minutes.
- To shape each portion into a long and narrow loaf-shape, apply the following process:
  - Flatten each portion into a rough oval shape, fold the edges of the dough towards the centre, and press gently to seal.
  - Then flip the dough over, so the seam side is facing down.
  - Finally, press and stretch the dough gently with your hands to elongate it, while maintaining an oval shape.
- Inside a dough ball tray, transfer the shaped-loaves onto a proofing couche cloth or a floured or parchment-lined baking sheet, keeping adequate space between them, which should be covered and allowed to rest for about 1 hour for the loaves to rise before baking.
- Load the rested dough on a right-size peel paddle (which has to be lightly floured), spray water onto the dough, and then skore the dough lengthwise with a razor at a slight angle. Note that the razor (the bread scoring knife) can be greased with butter or dipped into the water to make an easy cut.
- Bake the loaves for 20–25 minutes when the wood-fired oven temperature is about 200–220°C and without any flame-radiated heat until they are golden brown and sound hollow when tapped on the bottom.
- Let the loaves cool on a wire rack before slicing and serving.

7.3.2.3 Hints and Comments

Turkish loaf, also known as "Somun ekmek" and simply "Ekmek" in Turkish, is a type of bread that is commonly consumed almost in every meal in Turkish cuisine. It is similar to a French baguette or Italian ciabatta, with a slightly chewy texture and a golden-brown crust.

It is often served with a variety of toppings, such as olive spread, olive oil, butter, jam, honey, cheese, or cold meats. In recent years, the half of the Turkish loaf forms the sandwich bread in popular street foods, kebabs, fish, doner, and kokorec.

The dough can also be used as a base for traditional flat breads, "pide" and "lahmacun." In some versions of the bread, cooked potato is added to the dough.

7.3.3 Simit (Turkish Bagel) (see Figure 7.2)

7.3.3.1 Ingredients

The 4 cups (500g) of all-purpose flour, 2 tsp (10g) of salt, 1 tsp (4g) of sugar, 1 tbsp (12g) of instant yeast, 1.25 cups (300ml) of warm water, 0.25 cup (60ml) of olive oil, 0.5 cup (120ml) of grape molasses diluted with 0.5 cup (120ml) of water, and 1.5 cups (250g) of sesame seeds.

"Istanbul Simit's" Ingredients: 1kg all-purpose flour, 500ml water, 20g fresh yeast, 25g butter, 30g sugar, 20g salt, 120ml molasses diluted with 120ml water, and 250g sesame seeds.

1 cup of sugar and 1.5 cups of water.

7.3.3.2 Preparation and Cooking

- In a large mixing bowl, combine the all-purpose flour, salt, sugar, and instant yeast. Mix well, and gradually add the warm water and olive oil to the dry ingredients. Mix until a dough forms. Note that the dough of simit will be denser than other dough products. Therefore, the mixing should be done manually or by a commercial-size dough mixer, as home mixers may fail. If a dough mixer is used, mix for 3 minutes at slow speed, add the yeast, and mix for 7 minutes more at a higher speed.
- Turn the dough out onto a lightly floured surface and knead for about 10 minutes, until it becomes smooth and elastic. Add more flour if necessary to prevent sticking.
- Shape the dough into a ball and place it in a lightly oiled bowl. Cover with a damp cloth or plastic wrap and let it rise in a warm place for about 1–1.5 hours, or until it has doubled in size.
- After the dough has risen, gently deflate it and divide it into eight equal pieces. Roll each piece into a long rope about 40 cm long. Fold each rope in half and twist the two strands together, then join the ends to form a circle. Press the ends firmly to seal.
- In a shallow dish, mix together the molasses and water. In another shallow dish, place the sesame seeds.
- Dip each shaped dough circle into the grape molasses–water mixture, making sure it's fully coated. Then, dip it into the sesame seeds, pressing gently to ensure they stick.
- Place the coated dough circles on a parchment-lined baking sheet, allowing enough space between them for rising. Cover with a damp cloth or plastic wrap and let them rise for about 30 minutes, or until slightly puffy.
- As the wood-fired oven temperature is around 200–220°C, bake the simit for 20–25 minutes, or until they are golden brown and crusty. Note that it may be necessary to rotate the simits halfway through for even browning.
- Remove the simit from the oven and let them cool on a wire rack.

FIGURE 7.2   Stages of Simit (Turkish Bagel): Twisted dough ropes, formed into circular shapes and partially cooked simits.

Ertuğrul, Nesimi, Personal photograph.

- Simits can be served as fresh or stored in an airtight container for up to two days or can be frozen in a freezer for a longer period to reheat before serving.

7.3.3.3 Hints and Comments

Simit, also known as the Turkish bagel, is a circular bread encrusted with sesame seeds. It is a highly popular breakfast and street food in Türkiye and has a delicious and slightly chewy texture.

7.3.4 Black Sea Pide, Closed-Type Pide (Samsun Pide) (see Figure 7.3)

7.3.4.1 Fillings, Option 1

500g of minced lamb, beef, or goat meat (with 20–30% fat), 4 large size of onions, 4 medium-sized tomatoes, 1 green bell pepper, 1 tbsp of tomato paste, 1 tbsp of pepper paste (mild or hot), 2 tsp ground cumin, 1 tsp paprika, salt and black pepper to taste, 1 bunch of fresh parsley, 1 egg, and 1/4 cup of butter.

7.3.4.2 Fillings, Option 2

500g of minced lamb, veal, or beef (with 20–30% fat), 1 medium-sized onion, 1 large red capsicum, or 2 medium-sized tomatoes, 200p of mushrooms, a bunch of parsley, 2 cloves of garlic, ¼ cup of olive oil, 1 tbs of tomato paste, 1 tbs of red pepper paste, 1 tsp of salt, 1 tsp of black pepper, 1 tsp of paprika, 1 egg, and 1/4 cup of butter.

7.3.4.3 Preparation and Cooking

- In Samsun Pide, unlike the open-pide versions, before rolling-out the dough, all the ingredients have to be pre-cooked and ready for filling.
- Preparation of the fillings: in a large pan in the wood-fired oven, cook finely chopped onion using olive oil, then add minced meat and cook until lightly cooked. Then add the remaining ingredients after finely chopping the tomatoes, garlic, and fresh peppers, season the mixture, and cook it for about 20 minutes.
- Divide the pide dough into equal portions and roll-out the dough longer and thinner than the open-pide types.
- Place the pre-cooked fillings on it generously, leaving a 2 cm border around the edges. Fold the long sides of the rolled-out dough towards the centre and pinch them together to seal firmly. Note that unsealed dough will allow the escape of steam as well as the spread of ingredients over the oven floor.
- Brush the top of the closed-pide with egg wash and transfer it inside the wood-fired oven while leaving some space between them since they will expand during baking and extra room is needed to operate long-thin peel paddles to change the orientation of pides.
- Bake the pides in the oven until golden and crispy in Area E while moving the pide in and out closed to Area D for uniform cooking. Note that typical cooking time may vary around 10–15 minutes and at around 270°C, or until the top of the pide is golden brown.
- After cooking, remove the Samsun Pide from the oven and immediately brush with melted butter (which enhances the flavour and texture of the pide) before slicing into approximately 4 cm-thick pieces using a wheel cutter or a knife.

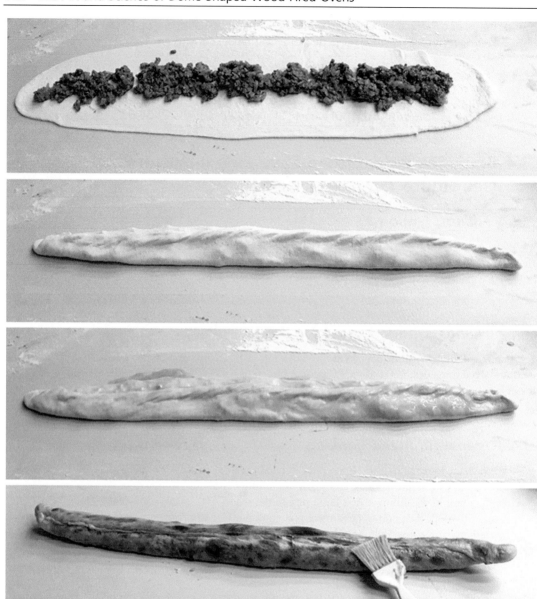

FIGURE 7.3   Closed-type Pide, Samsun Pide: Thinly rolled-out dough with fillings, folded and egg-washed pide, and freshly baked pide brushed with butter.

Ertuğrul, Nesimi, Personal photograph.

7.3.4.4 Hints and Comments

- This type of pide is served as a main dish, typically after cooling down to allow the fillings to cool as well. It can be accompanied by side dishes such as pickles, a fresh salad, yogurt drink (ayran), or spicy purple carrot pickle juice (Şalgam Juice).

- The size of the rolled-out dough and the amount of filling inside each closed-pide will determine the required amount of egg and butter.
- Note that the filling options presented here have similar ingredients, with Option 1 featuring onion and tomato as the main ingredients.
- Melted butter gives the pide a richer taste and a slightly different aroma compared to olive oil, which may also be used during the final wash.

7.3.5 Onion-Based Lahmacun (Soganlı Lahmacun) (see Figure 7.4)

7.3.5.1 Fillings

500g of finely diced lamb, veal (or beef) with 20–30% fat (suet*) content; 3 large onions; 1 large red bell pepper or red capsicum; 2–3 banana capsicums; 2 medium-sized tomatoes; a bunch of parsley; 2 cloves of garlic; 1/4 cup of olive oil; 1 tbsp of tomato paste; 1 tbsp of red pepper paste; 1 tsp of salt; 1 tsp of black pepper; and 1 tsp of paprika (sweet, mild, or hot, optional).

7.3.5.2 Preparation and Cooking

- The dough for lahmacun is prepared using the recipe and instructions provided earlier for "plain pide bread." However, lahmacun is distinguished from pide by its thin texture and much smaller dough portion size. It has a round, uniformly flat shape with a diameter between 8 and 20 cm and 2–3 mm of thickness. Therefore, the typical dough portion size for lahmacun is less than the size of an egg.
- Preparation of the toppings: traditionally, lahmacun meat (lamb or beef) is diced very finely, down to the size of lentils. Additionally, the remaining vegetables, including onions, red bell peppers, green bell peppers, tomatoes, garlic cloves, and parsley (with stems and leaves), are very finely chopped separately. Then, the remaining ingredients, such as olive oil, pastes, ground cumin, paprika, salt, and pepper, are mixed well to combine. It is important to note that the texture of the filling is crucial for it to stick to the rolled-out thin dough, which should be approximately 2–3mm thick. The desired texture can be achieved by further mixing and by adding water or olive oil and/or pepper paste as needed.
- Roll each portion of dough out into a thin (2–3 mm thick) round shape with a diameter of 8–20cm.
- Spread a thin layer of the mixture onto each piece of dough. Note that there is no need to leave a border around the edges.
- The cooking time for lahmacun is about 5–8 minutes under the oven temperature of 250–300°C, or until the edges are crispy and the topping is cooked thoroughly.
- The lahmacun is frequently served hot and primarily garnished with a range of vegetables as given in Chapter 6. Although garnish mixtures can vary based on individual preference, the traditional garnish for lahmacun is fresh parsley and a squeeze of fresh lemon juice.

7.3.5.3 Hints and Comments

- "Lahmacun" can be translated as "meat with dough." The name comes from the Arabic "Lahm b'ajin," where "lahm" means "meat" and "'ajin" means "dough." However, its origin is likely linked to the historical flatbread products dating back to the Mesopotamian, ancient Anatolian, and East Asian civilizations.
- Lahmacun is a highly versatile food that can be stored in the freezer. When packing for storage, place a pair of lahmacun face-to-face, with the filled sides touching each other.

FIGURE 7.4   Onion-based Lahmacun (Soganlı Lahmacun): Pide dough portions ready for lahmacun, and the finished lahmacun, half-cut, with cucumber filling and a slice of lemon.

Ertuğrul, Nesimi, Personal photograph.

- It can also be served with Shish Kebabs.
- *Suet is the raw, hard fat found around the loins and kidneys of animals, particularly cows and sheep. Its high smoke point makes it ideal for pastry production and frying.

7.3.6 Wrapped Grape Vine Leaves (Asma Yapragi Sarmasi) (see Figure 7.5)

7.3.6.1 Ingredients 1 (with Meat)

1/2 kg of grape vine (and/or mulberry) leaves, 500g of minced lamb or beef (with 20% fat), 1/2 cup of rice; 1 onion; 1/4 cup of parsley; 1/4 cup of dill; 1 tbsp of tomato paste; 1 tsp of salt; 1/2 tsp of black pepper, 1/4 tsp of cumin; 1/4 tsp of paprika; 1/4 tsp of allspice; 1/4 tsp of cinnamon; 1/2 tbsp of sumac; 1 or 2 lemon (optional, sliced); and 3.5 cups of boiling water (or 1 cup of boiling water and 2 cups of fresh tomato sauce).

7.3.6.2 Ingredients 2 (No Meat, with Olive Oil)

1/2 kg of grape vine (and/or mulberry) leaves; 7 large onions; 1 cup of olive oil, 2 cups of coarse bulgur; 2 tbs of tomato paste; 3 tbsp of currants; 1/4 cup of pinenuts; 1/4 cup of parsley; 1/4 cup of dill; 1 tbs of mint; 1 tbs of basil; 2 tsp of paprika; 1 tsp of black pepper; 2 tsp of salt; pomegranate molasses and lemon juice (optional); and 3.5 cups of boiling water (or 1 cup of boiling water and 2 cups of fresh tomato sauce).

7.3.6.3 Preparation and Cooking

- Blanch the cleaned leaves in boiling water for 2–3 minutes, then pat them dry and trim the stems (if necessary).
- Sauté minced meat and chopped onions in olive oil until golden. Add tomato paste and cook for 5 minutes. Add rice (or bulgur) and water, and cook until softened. Add spices, herbs, salt, and nuts (if using).
- Place a leaf on a clean surface with the shiny side down. Add 1–2 tsp of filling near the stem edge.
- Fold the stem end over the filling, followed by the left and right sides. Roll the leaf away from you to form a tight cylindrical parcel.
- Line the bottom of a clay pot with extra vine leaves or lemon slices. Place the rolled leaves seam-side down in a single layer. Pour water (or tomato sauce) and olive oil over the pot, covering the leaves. Place additional leaves or a heatproof plate on top.
- Close the lid and place the pot in the wood-fired oven in Area C, E, or H.
- Cook at 170–190°C for 1–2 hours.
- Cool the wrapped leaves in the pot, then transfer to a serving platter and serve with yogurt sauce.

7.3.6.4 Hints and Comments

- The taste of leaves varies, with young grape vine leaves having a mildly tart and tangy flavour, young mulberry leaves being sweeter with a hint of bitterness, and young cherry leaves having a subtle, aromatic flavour.
- Cook wrapped leaves in water or fresh tomato sauce with garlic and/or lemon slices, based on preference.
- Layer leaves at the bottom and on top of the wrapped leaves to prevent unwrapping.

FIGURE 7.5　Wrapped Grape Vine Leaves (Asma Yaprağı Sarması): Grape leaves ready for picking, rich fillings, and rolled leaves prepared for cooking.

Ertuğrul, Nesimi, Personal photograph.

- Use larger vine leaves, but smaller leaves can be overlapped.
- For olive oil-based dishes, use pomegranate molasses or lemon juice. For meat-filled dishes, add sumac or sumac juice in the proper amount.

7.3.7 Kibbeh with Yogurt Sauce (İçli Köfte) (see Figure 7.6)

7.3.7.1 Ingredients 1

7.3.7.1.1 For Shell
One cup of fine bulgur; 2 tbsp of wheat flour; 1 cup of hot water (or more if needed); 0.5kg of minced beef or lamb divided into two portions; 1 medium onion; 1/4 cup of fresh parsley; 1/4 cup of green onions (optional); 1 tsp of ground cumin; 1 tsp of ground allspice; 1/2 tsp of black pepper; 1/2 tsp of paprika; and salt.

7.3.7.1.2 Filling
Two tbsp of olive oil; 0.5kg of minced beef or lamb; 1 small onion; 1/4 cup of fresh parsley; 1/4 cup of crushed walnuts (and/or pine nuts); 1/2 tsp of cumin; 1/2 tsp of allspice; 1/4 tsp of black pepper; and salt.

7.3.7.1.3 For Yogurt Sauce
Four cups of yogurt (preferably thick texture Turkish or Greek-style); 1 cup of water; 1 tbsp of corn-starch (or wheat or chickpea flour); 3 cloves of minced garlic; 2 tbsp of dried mint; and salt to taste.

7.3.7.2 Preparation and Cooking

- Soak fine bulgur in hot water for 20–30 minutes until softened.
- In a skillet, cook chopped onion in olive oil until softened. Add minced meat and cook until browned and cooked through. Stir in parsley, nuts, spices, and salt. Set aside to cool.
- Combine softened bulgur, meat mixture, onion, parsley, green onions (optional), and spices. Knead until well combined and smooth.
- Shape the kibbeh by forming small walnut-sized balls, making an indentation, filling with the prepared mixture, and shaping into an elongated oval or football shape.
- Repeat the shaping process with the remaining mixture and set aside.
- Whisk yogurt, water, and cornstarch in a pot until smooth. Cook over medium heat, stirring constantly, until slightly thickened.
- Place shaped kibbeh in the simmering yogurt sauce in a pot or tray. Simmer in a wood-fired oven at 150–170°C for 20–25 minutes.
- In a frying pan, sauté minced garlic and dried mint for 2 minutes. Serve kibbeh with yogurt sauce, rice, bulgur pilaf, pita bread, tahini sauce, or a salad.

7.3.7.3 Hints and Comments

- The Turkish version of this dish is deep-fried, while variations exist in Middle Eastern countries.
- In a wood-fired oven, baking or cooking in water or yogurt sauce is recommended. This version, known as kibbeh labanieh or kibbeh in yogurt sauce, is popular in Lebanese and Syrian cuisine.

7.3.8 Adana Shish Kebab (see Figure 7.7)

7.3.8.1 Ingredients

500g of minced lamb or veal or beef (or a combination of both), with 15–20% fat content; 1 medium onion (grated); 1 large capsicum; 1/4 cup of fresh parsley; 2 cloves of garlic; 1 tbsp of Turkish red pepper

FIGURE 7.6   Kibbeh with Yogurt Sauce (İçli Köfte): Combined mixture, kibbeh shaped into elongated ovals, and the serving presentation.

Ertuğrul, Nesimi, Personal photograph.

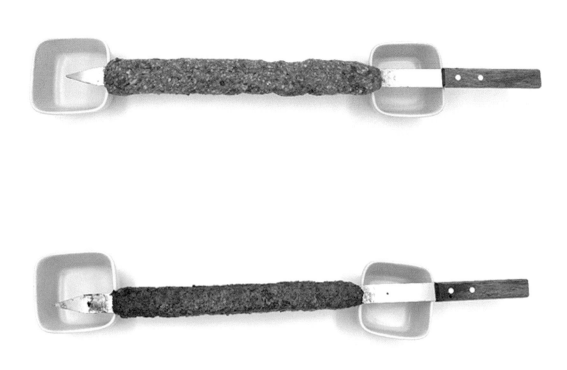

FIGURE 7.7 Adana Shish Kebab: Raw ingredients shaped around skewers and their appearance post-cooking.

Ertuğrul, Nesimi, Personal photograph.

flakes (pul biber); 1 tsp of ground cumin; 1 tsp of sumac (optional); 1 tsp salt; 1/2 tsp of black pepper; and metal skewers of about 30–50cm length.

7.3.8.2 *Preparation and Cooking*

- In a bowl, combine the ground meat, grated onion, finely chopped parsley, minced garlic, finely chopped capsicum (after squeezing to remove the juice of capsicum), red pepper flakes, cumin, sumac (if using), salt, and black pepper, and mix well with your hands until the ingredients are evenly distributed and the mixture is well combined.
- Divide the meat mixture into equal portions for each skewer. Shape each portion into a long (in the middle of the skewer and about 20–30cm long), flat, sausage-like shape around the skewers, pressing the meat firmly onto the skewer. Use icy cold water while loading the meat mixture. The final shape is a long sort of light wave-shaped pattie. Make sure to press the meat firmly onto the skewers to prevent them from falling off during cooking
- To cook the kebab, use a tray as described in the previous shish recipes. Note that for cooking, the skewers should not be placed directly on the oven floor of the wood-fired oven. Cook the shishes for about 20 minutes at 170–200°C.

- Keep an eye on them, as cooking times may vary depending on the temperature of the oven, the directions of the heat sources, and the thickness of the kebabs.
- To serve, slide the kebabs off the skewers directly onto flatbread (or pide bread) or on a serving plate. Accompany the Adana Kebabs with grilled vegetables, a fresh salad, yogurt, or a side of rice or bulgur pilaf.

7.3.8.3 Hints and Comments

- The key to cooking in a wood-fired oven is to achieve the desired doneness and texture.
- Use rib meat for its texture and fat content, and keep minced meat cold before shaping by refrigerating prior to mixing. In addition, refrigerate the mixture for about 30 minutes.
- Rub meat fat on skewers before loading the mixture to prevent sticking.
- To load the mixture, form a small ball and load it from the tip to the end of the skewer while extending and flipping back and forth. Use icy water to extend the mixture and press firmly on each end before cooking.
- Keep in mind that shish can sweat during cooking, so it is advised to remove it with flatbread or plain pide bread after cooking.

7.3.9 Clay Pot Kebab/Casarole (Güveç Kebab or Güveç) (see Figure 7.8)

7.3.9.1 Ingredients

Optional: 800g lamb or beef (cut into 2–3cm cubes); 2 medium onions; 2 bell peppers (various colour); 2 medium tomatoes; 2 bell peppers; 3 cloves of garlic; 1/4 cup of olive oil; 1/4 cup of tomato paste; 1/2 tsp of black pepper; 1 tsp of paprika; 1/2 tsp of cumin; 1/2 tsp of dried oregano; salt to taste; 1 cup of water or beef/lamb broth; further optional ingredients: potatoes, carrots, ladyfingers, chopped fresh green beans, and fresh parsley for garnish.

7.3.9.2 Preparation and Cooking

- Fully submerge the clay pot (both bottom and lid) in water for up to 1 hour.
- If preferred, the lamb cubes can be sealed on all sides briefly in olive oil. Combine the cubed meat, chopped onions, bell peppers, tomatoes, minced garlic, olive oil, tomato paste, black pepper, paprika, cumin, oregano, salt, and optional ingredients and mix all gently.
- Transfer the mixture to the clay pot and add the water or broth over the mixture, ensuring all ingredients are covered, 1–2cm below the top level of the pot. Then close the lid and make sure that the clay pot is sealed properly to retain moisture during cooking.
- Place the clay pot in Area E (or Area G) of the wood-fired oven and cook for about 2 hours at around 170–190°C.
- Using a metal peel paddle and a mitten, carefully remove the güveç from the oven and let it rest for 10 minutes before serving, or rest it in Area F to keep warm until serving.
- Serve the kebab hot directly from the pot, garnished with chopped parsley if desired. Pair it with warm flatbread, pide, loaf, or serve it over rice or bulgur pilaf.

7.3.9.3 Hints and Comments

- This "one-pot dish" features slow-cooked meat and there are variations depending on the region and personal preferences, which include additional herbs, spices, or vegetables, and some versions may use chicken or other types of meat.

FIGURE 7.8   Clay Pot Kebab/Casserole (Güveç Kebab or Güveç): Vegetable casserole in a clay pot, with stages of flat green beans and onion-based ingredients, leading up to the cooked form.

Ertuğrul, Nesimi, Personal photograph.

- Note that preparing a clay pot for cooking, often referred to as seasoning or curing the pot, is essential to prevent it from cracking during cooking. Seasoning involves soaking in water, oiling, curing in an oven, cooling, and repeating the same process. As stated in the previous chapter, after cooking, clean the pot by hand with warm water and a soft brush, avoiding using detergent.
- As shown in the photo, the dish is also prepared as an all-vegetable casserole and has a Turkish name "Turlu" that has the following typical ingredients: aubergine, zucchini, sweet green peppers, okra, green beans, tomatoes, olive oil, onions, garlic, and parsley.

7.3.10 Tepsi Kebabi (see Figure 7.9)

7.3.10.1 Ingredients

800gr minced lamb, 150gr caul or tail fat, 1 medium onion (finely chopped), 3 garlic cloves (minced), 2 red bell peppers (finely chopped), 2 banana capsicum (finely chopped, and use the other 4 for decoration), 1 bunch of parsley (finely chopped), 2 medium tomatoes (sliced in quarter for decoration), 1 tsp of salt, 1 tsp of red pepper or chilli powder, 1 tsp of black pepper, 1 tsp of cumin, and 1 tbsp of butter.

7.3.10.2 Preparation and Cooking

- In a bowl, mix the minced lamb and caul or tail fat together, and add the salt, red pepper or chilli powder, black pepper, and cumin to the mixture, and mix thoroughly.
- Then add the finely chopped onion, garlic, red bell peppers, banana capsicums, and parsley to the mixture.
- Grease a round pan (similar to a paella pan or skillet or a shallow earthenware dish) with butter.
- Spread the lamb mixture evenly in the pan, applying medium pressure to make sure it covers the entire surface, and slice the mixture like a pie, making 8 or 16 slices based on the size of the pan.
- Decorate the top of the mixture with the remaining banana capsicums and sliced tomatoes.
- In a separate bowl, mix 1 cup of water and 1 tablespoon of tomato paste to make a sauce. Pour the sauce over the top of the kebab.
- Cook the kebab in a wood-fired oven at a temperature of 160–200°C for about 30 minutes in Area E or less in Area C.
- After cooking, the kebab can be kept in Area F while waiting for service.
- Tepsi Kebab can be served warm, as in other kebab and shish options, with yogurt drink or pickled red carrot or beetroot juice (known as Şalgam Suyu).

7.3.11 Slow-Cooked Goat (Fırında Keçi) (see Figure 7.10)

7.3.11.1 Ingredients

Whole goat (8–15 kg, cleaned and cut into larger pieces); 1/2 cup of olive oil; salt and pepper, rosemary, thyme; lemon zest and juice; small onions (optional); 20–30 cloves of garlic (optional); and a bottle of white or red wine (optional).

1 tbsp of cumin; 1 tbsp of paprika; 1 tbsp of coriander; 1tbsp of turmeric; 1 cup chicken broth; and 1 tbsp of olive oil.

7.3.11.2 Preparation and Cooking

- Using a correct-size stainless steel tray (to fit into the door cavity), pack the pieces of goat after rubbing each piece with olive oil and seasoning liberally with salt, pepper, and preferred

FIGURE 7.9 Tepsi Kebabi: A spread of lamb mixture in a pan, adorned with fresh banana capsicum and tomato slices, and the dish after cooking.

Ertuğrul, Nesimi, Personal photograph.

FIGURE 7.10 Slow-Cooked Goat (Fırında Keçi): Images of goat meat slow-cooked for 14 hours in stainless steel trays.

Ertuğrul, Nesimi, Personal photograph.

herbs and spices. If desired, lemon zest and juice and garlic cloves can be used for added flavour. If time and space permit, the meat can be left to marinate overnight.

- Add 1cm of hot water (and wine, if preferred) to the tray, cover tightly with aluminium foil, and place it on the Area G of the oven floor. Then in the slow cooking mode of the oven (with no active fire and the door closed), cook it for about 10–12 hours at a temperature of 120–140°C.
- After such a cooking process, the meat becomes tender and is pulled off the bone, while a reasonable amount of stock (due to the long duration and thickness of the liquid) is collected at the bottom of the tray. Remove the stock from the tray which can be used for highly flavoursome substitute in many dishes. If the oven is re-fired for further cooking, roast the cooked goat in Area A or Area C for 5–10 minutes which will enrich the flavour with a crispy surface.
- Slow-cooked goat can be served in a variety of ways:
  - As a cold appetizer after slicing into bite-sized pieces and serving as a starter with pickles, olive oils, and pide bread on the side;
  - In a soup using broth to prepare a hearty soup including carrots, onions, legumes, salt, pepper, and herbs, accompanied by pide or loaf bread;
  - In "Iskender Kebab," after thinly slicing the goat meat;
  - Slicing the rest of the meat into bite sizes, making single-serve packaging, and storing in the freezer.
  - Traditional Turkish drinks, like ayran (a yogurt-based drink) or Şalgam (a fermented red turnip juice), are great accompaniments to the rich slow-cooked goat dishes listed above.

7.3.11.3 Hints and Comments

- Goat meat can be tough, which makes it a less popular choice in many cuisines. However, slow cooking in a wood-fired oven not only tenderizes this meat to perfection, but also amplifies its flavour. Note that various game birds can also be slow-cooked following this recipe, including pheasant, quail, duck, partridge, guinea fowl, and wild turkey. The end result is a dish featuring tender and flavourful meat.

7.3.12 Stuffed Pumpkin/Fusion (see Figure 7.11)

7.3.12.1 Ingredients 1 (with Meat)

One butternut or buttercup pumpkin (about 1.5–2kg); 200g of slow-cooked and finely chopped goat meat (or beef or lamb or chicken); 1 cup of coarse bulgur (or rice, soaked in water for an hour); 1 onion; 4 tbsp of olive oil; 1 tbsp of tomato paste; 1 tsp of salt; 1/2 tsp of black pepper; 1/4 tsp of cumin; 1/4 tsp of allspice; and 1/4 tsp of cinnamon.

7.3.12.2 Ingredients 2 (without Meat)

One butternut or buttercup pumpkin (about 1.5–2kg); 2 large onions; 1/2 cup of olive oil; 1 cup of coarse bulgur (or rice, soaked in water for an hour); 1 tbs of tomato paste; 3 tbsp of currants; 1/4 cup of pinenuts; 1/4 cup of parsley; 1/4 cup of dill; 1 tbs of mint; 1 tbs of basil; 2 tsp of paprika; 1 tsp of black pepper; 1.5 tsp of salt; pomegranate molasses and lemon juice (optional); and 2 cups of boiling water (or 1 cup of boiling water and 1 cup of fresh tomato sauce); if available 1/4 cup of half cooked and diced chestnuts (optional); 1/4 of finely chopped dry apricot and date (optional).

7.3.12.3 Preparation and Cooking

- Cut off the top of the pumpkin to create a lid, and scoop out the seeds and fibrous material inside the entire pumpkin using a spoon, leaving the harder shell section.

FIGURE 7.11 Stuffed Pumpkin/Fusion: Coarse bulgur-based ingredients, a filled and carved pumpkin, the pumpkin with its lid pre-cooking, and the post-cooking presentation.

Ertuğrul, Nesimi, Personal photograph.

- Prepare the choice of ingredients given above following the instructions given previously under stuffed and wrapped dishes.
- Fill the hollow pumpkin with the stuffing mixture and replace the lid of the pumpkin.
- Bake the pumpkin in the wood-fired oven at around 150–180°C for about 1–2 hours or until the skin of the pumpkin is gently burnt.
- When the pumpkin is done, remove it from the oven and allow it to rest for only a few minutes before serving. You may cut it into slices or serve the stuffing separately, but it is recommended to serve directly from the pumpkin by scooping out a mixture of flesh and the cooked fillings.

7.3.12.4 Hints and Comments

- This dish can be considered as a one-pot cooking without a real clay pot and without clay pot preparation and dish washing.
- Note that due to the size of the pumpkin selected and the thickness of the flesh, an accurate measure of the stuffed ingredients may vary.
- This dish can be highly flavoursome and diverse since the final dish combines the smokey pumpkin and the selected range of ingredients. If a butternut pumpkin is selected, the cooked flesh becomes tender and offers a slightly sweet and buttery flavour, similar to sweet potatoes or caramelized roasted carrots. However, when buttercup pumpkin is used, its taste may be described as a richer and more intense sweetness and also offer a hint of nuttiness, akin to roasted chestnuts.
- A teacup of molasses can also be added to the recipe for additional flavour.
- Note that there is a Thanksgiving version of a stuffed pumpkin in the US, which can also be cooked in the wood-fired oven to enhance and introduce different flavours.

7.3.13 Water Börek (Su Böreği) (see Figure 7.12)

7.3.13.1 Ingredients

For the dough: 3.5 cups of flour; 1.5 tsp of salt; 6 eggs.
For boiling: 3–4 litres of hot water; 1 tsp of salt.
For the layers: 1/2 cup of olive oil; 150 grams of butter; milk and egg (optional for layers).
For the filling: 600 grams of white cheese (such as feta cheese).

7.3.13.2 Preparation and Cooking

- Begin by combining eggs, salt, and flour in a bowl and mixing them together using a wooden spoon. Keep kneading until you achieve a medium-soft dough that is not sticky.
- Next, shape the dough into small balls about the size of an orange, place them on a floured surface, cover with a cloth to proof, and let them rest for 30 minutes.
- Roll out the first ball of dough into a thin layer, similar to phyllo or yufka, ensuring it's large enough to overlap the edges of your baking dish.
- Boil each pastry sheet in a large pan filled with water (or milk) for approximately 30 seconds without stirring. This step imparts a unique, slightly chewy texture to the börek.
- Prepare your baking dish by greasing it with a mixture of melted butter and olive oil.
- After boiling, immediately immerse the pastry in ice-cold water to stop the cooking process, then gently squeeze out any excess water. Arrange the pastry in the baking dish, allowing it to take on a creased or wrinkled appearance.
- In a separate bowl, crumble the feta cheese (and add chopped parsley, if desired).
- Lightly brush each pastry layer with the butter and olive oil mixture.

FIGURE 7.12   Water Börek (Su Böreği): Egg-combined dough, dough rolled out and ready for boiling, and the final baked water börek ready for serving.

Ertuğrul, Nesimi, Personal photograph.

- When you've layered half of the pastries, distribute the crumbled cheese evenly over the last layer. Continue with the layering process until all the dough balls are used.
- After the final layer is in place, gently press down the top to even it out, and brush it with an egg/milk wash.
- Bake the dish in a wood-fired oven in Area E, without direct flame-radiated heat, for approximately 50–60 minutes at around 190–210°C. Please note that cooking times may vary depending on the number of layers and the overall thickness of your börek. After the first 10 minutes of baking, turn the börek to ensure even browning and continue baking until the top turns golden and crispy.
- Allow the börek to cool slightly before slicing it into squares or rectangles. It is best served warm. This börek is a versatile dish that can be enjoyed for breakfast, as a tea time snack, or even as a light lunch or dinner when paired with a fresh salad, a bowl of soup, or included in a mezze platter.

7.3.13.3 Hints and Comments

- Due to the laborious process of making Su Böreği, it is often a communal activity, bringing together family members or friends for an afternoon of cooking.

7.3.14 Baklava (see Figure 7.13)

7.3.14.1 Ingredients

About 40 sheets of thin phyllo pastry (about 30cm × 30cm each), ¾ cup of sugar, 1 ¼ cup of water, 1 tbsp of lemon juice, 300gr of ghee (or clarrified unsalted butter), and 300gr of crushed walnuts (or pistachios).

7.3.14.2 Preparation and Cooking

- To prepare the syrup, combine water and sugar in a saucepan. Bring this mixture to a boil over high heat and let it simmer for about 5 minutes. Introduce the lemon juice and continue to simmer for an additional 5 minutes until it begins to slightly thicken. Remove from heat and let it cool.
- Grease a baking tray (approximately 30cm × 30cm × 5cm) evenly with ghee.
- Begin assembling the baklava by first allowing the phyllo pastry to thaw in its sealed packaging at room temperature for roughly 2 hours. Then, arrange three sheets of phyllo pastry in the tray, leaving some space for air gaps. Brush or spray the layers with a mixture of ghee. Repeat this step about 5 times, leading to a total of 15 layers.
- Spread the crushed walnuts evenly across the arranged pastry layers, and continue the process of layering and brushing ghee until all the remaining phyllo sheets are used (approximately 20 layers). Ensure the final layer consists of a single sheet.
- Cut the baklava into your preferred shapes using a sharp knife. Use a ruler or wooden template as a guide to achieve uniform cuts. Clean the knife with a damp kitchen towel after each cut. For a neater look, trim any excess dough and tuck the uneven edges under the layers.
- Drizzle the remaining ghee mixture uniformly over the cut baklava.
- Mist the top layer with water before baking.
- When the temperature of the wood-fired oven is 180–200°C, bake the baklava for about 30 minutes, until the top layer is golden and crusty. Consider baking in the middle of Area E where there's no active fire, or at the onset of slow cooking in Area G.
- Upon removing the baklava from the oven, carefully retrace the cut lines to separate the pieces. Drizzle the cooled syrup evenly over the warm baklava and allow it to sit for a few hours to fully absorb the syrup.

FIGURE 7.13 Baklava Images: Cooked baklava, rolled and twisted variations, and Heavenly Mud (Cennet Çamuru).

Ertuğrul, Nesimi, Personal photograph top.

- Allow the baklava to cool for approximately 10 minutes so it can fully absorb the syrup before serving. You may serve it plain or with your favourite toppings or even can be served cold.

7.3.14.3 Hints and Comments

- This classic baklava recipe featured crushed walnuts, though variations may include pistachios, hazelnuts, almonds, or peanuts, and is often garnished with powdered pistachio and served in small servings with Turkish coffee, tea, or ice cream.
- For exceptional baklava, quality samna (ghee or clarified butter) and proper air gaps between phyllo layers are crucial. Ghee is preferred in baklava for its unique flavour and acidic taste and its resistance to burning. Homemade ghee can be made from ordinary butter through careful heating, chilling, and skimming processes.

7.3.15 Rolled and Twisted Baklava (Burma Baklava) (see Figure 7.13 middle)

7.3.15.1 Ingredients

A pack of the phyllo pastry (about 375g); 250g of ghee; 50g of sunflower oil (or 300g of butter to make ghee); 300g of finely crushed walnuts (or 400g of finely crushed pistachios); 3/4 cup of sugar, 1.2 cups of water, and 1 tbsp of lemon juice.

7.3.15.2 Preparation and Cooking

- Thaw the phyllo pastry for 1–2 hours and keep it under a damp cloth. Grease the bottom of a square baking tray with ghee.
- Brush ghee on a layer of phyllo and roll it twice around a wooden stick. Sprinkle walnuts on the unrolled section, leaving space at the end. Roll tightly and twist the ends to create a twisted appearance.
- Place the twisted phyllo in the baking dish and repeat the process until all phyllo and filling are used.
- Bake in Area E of the wood-fired oven at 200–280°C for 20–30 minutes until golden brown.
- Prepare the syrup by boiling water and sugar, then adding lemon juice and boiling for 5 more minutes.
- Let the syrup cool, then pour it over the warm baklava. Allow it to absorb the syrup for a few hours.
- Serve the baklava at room temperature. Enjoy it on its own or with Turkish coffee or ice cream. Garnish with chopped nuts, cinnamon, or powdered sugar.

7.3.16 Heavenly Mud (Cennet Çamuru) (see Figure 7.13 bottom)

7.3.16.1 Ingredients

300g kataifi; 125g of butter (or ghee); 1 cup of sugar; 1.2 cups of water; or 1.2 cups of milk (optional); 100g of kaymak (Turkish clotted cream); 3/4 cup of crushed walnuts (or pistachios); and 1 tbsp of lemon juice.

7.3.16.2 Preparation and Cooking

- Unravel kataifi (the shredded phyllo dough) and place it in a square-shaped baking tray (to achieve a kataifi mixture about 2.5cm thick) and spread the dough evenly.
- Melt the butter and pour it over the kataifi, and add the walnuts and mix, making sure all the strands are covered. Then level the mixture by pressing gently over the tray.

- Place the tray in Area E of the wood-fired oven and bake at 200–280°C without flame-radiated heat for 20–30 minutes, until the Heavenly Mud turns golden brown.
- While the dessert is baking, prepare the syrup. Combine water and sugar in a saucepan and bring it to a boil. After 5 minutes, stir in lemon juice and boil for another 5 minutes to lightly thicken the mixture.
- Slowly add the syrup to the hot kataifi and allow it to sit for a few hours to fully absorb the syrup.
- It can be served plain or with cream (kaymak) on top.
- This dessert can wait a few weeks in the fridge, but it should be heated and served warm.

7.3.17 Roasted Chestnut (Kestane Kebab) (see Figure 7.14)

- Select fresh chestnuts, making sure that they have a firm, bright, and shiny surface.
- Use a sharp knife to score with a cross on the round side, and soak in hot water or cold water if cooked the next day. There are three main purposes of scoring:

FIGURE 7.14   Roasted Chestnut (Kestane Kebab): Fresh chestnuts prepared for scoring and their appearance after roasting.

Ertuğrul, Nesimi, Personal photograph.

- to prevent them from exploding in the oven,
- to create an opening for the hot water to penetrate when soaked before cooking and loosen the shell and the inner skin, known as the pellicle, which makes it easier to peel,
- to allow them to hydrate, which can improve their texture when cooked, hence making them more tender and moist.

- Roasting chestnuts can be done in the wood-fired oven on a shallow oven-safe tray at a range of temperatures of 150–300°C in Area E or Area C for 15-20 minutes.
- After roasting, remove the chestnuts from the oven and let them cool for a few minutes until they are safe to handle. While still warm, peel off the outer shell and the inner skin to reveal the chestnut meat inside. Roasted chestnuts can be eaten as is.
- The chestnuts are a popular winter treat and can be prepared in various other ways, boiling or grilling.

7.3.18 Roasted Aubergines (see Figure 7.15)

- Roasting aubergines in a wood-fired oven is likely one of the most rewarding culinary practices. The skin becomes charred while the flesh turns incredibly tender, absorbing the smoky flavours from the wood fire, resulting in a distinctive taste and texture. This method is particularly favoured in Turkish and Middle Eastern cuisines.
- Choose fresh, firm, and brightly coloured medium- to large-sized aubergines, and wash and dry them. Then poke several holes all around using a fork to help steam escape during the roasting process.
- Rub the aubergines all over with olive oil.
- Place the aubergines in a sufficiently deep stainless-steel tray to prevent oil spills, and roast them at 220–270°C for about 30–40 minutes until the outer skin is charred and the aubergines have collapsed in on themselves. Note that aubergines have to be turned over once to ensure they are roasted/charred evenly.
- Once done, remove them from the oven and let them cool enough to handle. Be cautious that the olive oil will retain heat for a longer period of time.
- Slice the aubergines lengthwise to open the charred skin and scoop out the soft flesh using a dough scraper or a spoon, then chop all the flesh and add salt to taste.
- The final roasted aubergines can be served with yogurt mixed with dill, paprika, crushed garlic, and, if preferred, drizzled with the leftover juice and olive oil from the roasting process.

7.3.19 Roasted Capsicums (see Figure 7.16)

- Roasting capsicums in a wood-fired oven can be done similarly to aubergines.
- Once the outer skin is charred and blistered, let them cool until safe to handle, then peel off the skins, remove the stems and seeds, slice the flesh into short strips, and add salt to taste. Note that the capsicums need to be turned over once during roasting to ensure they char evenly.
- The roasted capsicums can be combined with finely chopped pickles like gherkin, cucumber, or cabbage and served with a yogurt dressing mixed with dill and paprika. For an added burst of flavour, drizzle the mixture with leftover juice and olive oil from the roasting process.
- In addition, there are various ways to serve roasted capsicums:
  - Can be added to salads with feta cheese, olives, and a simple vinaigrette.
  - Can be layered onto sandwiches or wraps.
  - Can make a great topping for pasta dishes when mixed with olive oil, garlic, and parmesan cheese.
  - Can be a side dish, after sprinkling with fresh herbs.

FIGURE 7.15 Roasted Aubergines: Fully roasted aubergines with charred skin, the soft interior revealed after skin removal, and the chopped flesh ready to serve alone or with accompanying dishes.

Ertuğrul, Nesimi, Personal photograph.

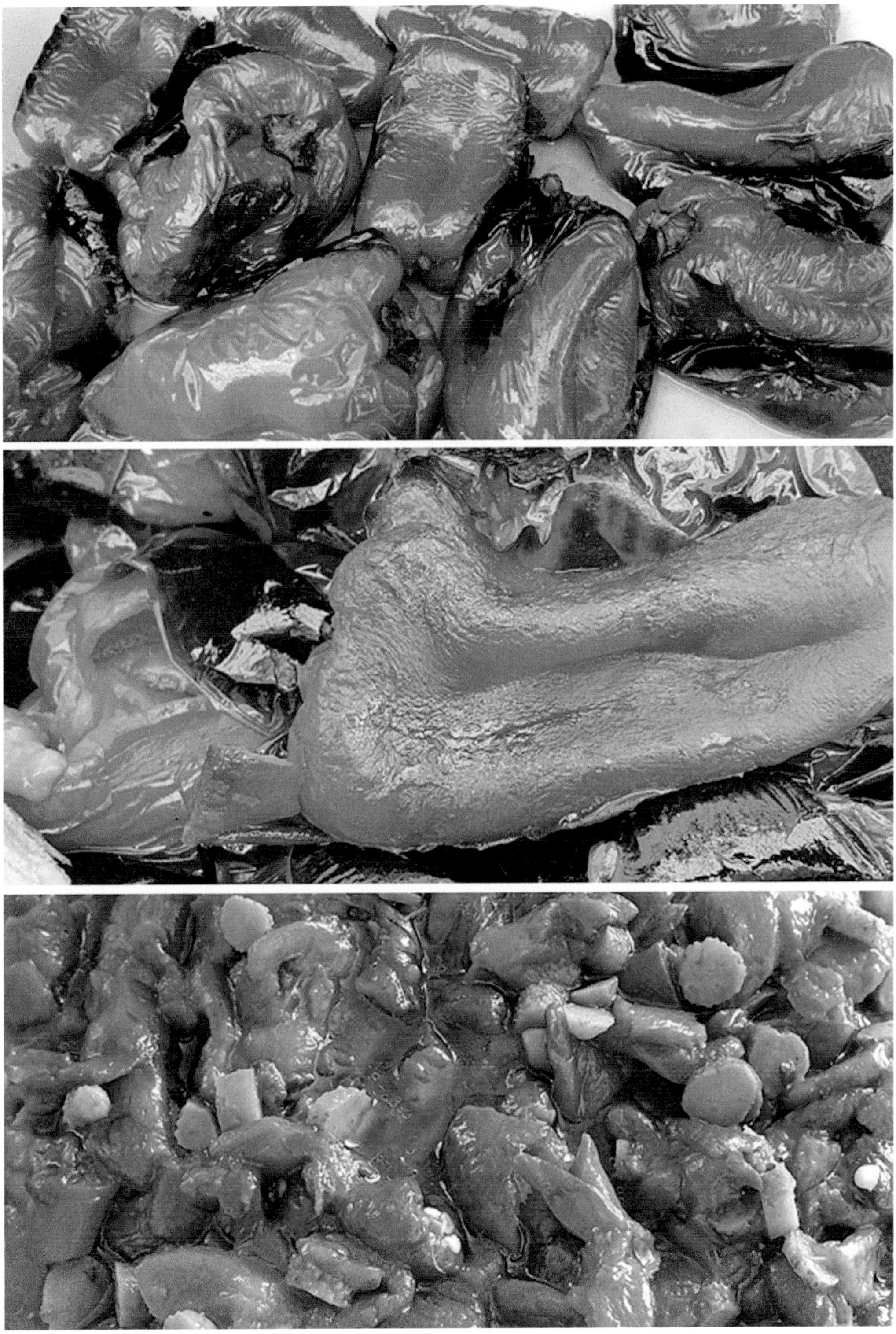

FIGURE 7.16 Roasted Capsicums: Roasted capsicums shown post charring, the flesh after removing the charred skin, and the chopped flesh mixed with finely diced gherkin.

Ertuğrul, Nesimi, Personal photograph.

FIGURE 7.17   Ruby Elixir/Fusion: A halved pomegranate displaying its seeds, the arils processed to extract juice, and the final cocktail served in a glass.

Ertuğrul, Nesimi, Personal photograph.

- Can be combined with salsa or used as a topping for grilled meats.
- Alternatively, it can be served on a slice of crusty fresh bread topped with goat cheese as a snack.

7.3.20 Ruby Elixir/Fusion (see Figure 7.17)

- The primary ingredients of this invigorating cocktail are freshly prepared pomegranate juice (70%) and vodka (or tequila, pisco, or rum) of your choice (30%). It's enhanced with fresh herbs such as mint and basil, a touch of black pepper, a touch of wasabi paste, very finely chopped young lemon leaves (optional), and a splash of lemon juice (optional).
- To prepare fresh pomegranate juice:
  - Cut off the crown (the protruding blossom end) of the pomegranate, taking care not to pierce the seeds.
  - Lightly score the skin in quarters from top to bottom, taking care not to cut into the seeds.
  - Over a large bowl, pull apart the quarters and gently pry the seeds (known as arils) from the pith and skin, keeping the seeds intact to prevent bursting.
  - Place the arils in a zip-top bag, seal it, and roll it with a rolling pin to burst the arils and release the juice. Cut a corner off the bag and squeeze the juice into a bowl or jug. Alternatively, a blender or mixer can be used with a pulsing action for about a minute in total to break open the arils and release the juice. Avoid overblending, as this can crush the inner pit and add a bitter taste to the juice.
  - Finally, strain the mixture through a fine-mesh strainer or cheesecloth to remove the crushed seeds and pits.
- For serving:
  - Choose your preferred herb and finely chop it. The pieces should be small enough to adhere to the glass rim but large enough to impart flavour.
  - Moisten the glass rim by running a lemon or lime wedge around it.
  - Spread the chopped herbs out on a flat surface or plate. Invert the glass and dip the moistened rim into the herbs, twisting the glass to ensure the rim is thoroughly covered.
  - Fill the glass with the fresh (ideally chilled) pomegranate juice and chosen alcohol, garnishing with a sprinkle of freshly ground black pepper to finish.

REFERENCES

[1] *Bread: A Baker's Book of Techniques and Recipes*, by Jeffrey Hamelman, Wiley, 2012, ISBN: 978-1118132715.
[2] *The Bread Baker's Apprentice: Mastering the Art of Extraordinary Bread*, by Peter Reinhart, Ten Speed Press, September 6, 2016, ISBN-13: 978-1607748656.

Index

Note: Locators in *italics* represent figures and **bold** indicate tables in the text.